THE STATE AND THE INTERNATIONAL OIL MARKET

Studies in Industrial Organization

Volume 23

The titles published in this series are listed at the end of this volume.

THE STATE AND THE INTERNATIONAL OIL MARKET

Competition and the Changing Ownership of Crude Oil Assets

Coby van der Linde
Netherlands Institute of International Relations 'Clingendael' and Leiden University, The Netherlands

Springer Science+Business Media, LLC

Library of Congress Cataloging-in-Publication Data
Linde, Coby van der.
The state and the international oil market : competition and the changing ownership of crude oil assets / Coby van der Linde.
p.cm. -- (Studies in industrial organization ; Bd. 23)
Includes bibliographical references and index.
ISBN 978-1-4613-7066-6 ISBN 978-1-4615-4575-0 (eBook)
DOI 10.1007/978-1-4615-4575-0
1. Petroleum industry and trade--Government policy. I. Title. II. Series.

HD9560.6 .L55 1999
382'.42282--dc21 99-048097

Originally published by Kluwer Academic Publishers in 2000
Softcover reprint of the hardcover 1st edition 2000

Printed on acid-free paper.

TABLE OF CONTENT

PREFACE

This book has been kicking around my desk for quite some time. On and off I returned to my work on the role of the state in the economy and the international oil market, but for a long time I was not satisfied with the shape it was in. I understand now that I needed the insights developed over the past couple of years on the role of the state, regulation, liberalization, privatization, and the recent events in the international oil industry to bring all my ideas together in a more coherent format. It was the events that followed the Asian financial crisis that drew me back to finish writing this book.

The early beginnings of this book were developed at the Institute of International Affairs, Chatham House, in London, where I was a research fellow with the Energy and Environment Programme in 1992 and 1993. At the Colorado School of Mines, I had the opportunity to test my ideas in a graduate class, and continue the research. In the years following, I published some of my ideas and worked on (parts of) the book at irregular intervals. Teaching and other interesting areas of research, however, kept me from finishing it. Only when I was appointed senior research fellow at the Institute for International Relations 'Clingendael' in the Hague in 1998, did my determination to complete the book return.

I would like to thank both Chatham House and Clingendael for offering me the opportunity to pursue this research and to write. Also, I thank Leiden University for releasing me from teaching duties during these two periods so that I could concentrate fully on writing. I am sincerely grateful for the comments provided on earlier versions of the book by Prof. Dr. H.W. de Jong, my long time-mentor. His brief but painfully insightful comments on how to make an argument work are always motivating. I greatly value the friendship of him and his wife. I would also like to thank Prof. Richard Higgott and Prof. Hans Schenk for their encouragement. I greatly appreciate the motivating comments of Joe Mc Cahery.

My daughter Meagan I thank for her cheerful acceptance that she will not be going to Italy this summer.

The Hague, Summer 1999

ABBREVIATIONS

b/d	barrels a day
BP	British Petroleum
CEPE	Ecuadorian Government Petroleum Corporation
FDI	Foreign Direct Investment
GATT	General Agreement on Tariffs and Trade
GCC	Gulf Cooperation Council
GDP	Gross Domestic Product
IMF	International Monetary Fund
KPC	Kuwait Petroleum Company
LDC	Less Developed Countries
MNC	Multinational Corporation
NIOC	National Iranian Oil Company
NNPC	Nigerian National Petroleum Company
NOC	National Oil Company
OECD	Organization of Economic Co-operation and Development
OPEC	Organization of Petroleum Exporting Countries
PDVSA	Petroleos de Venezuela SA
Pemex	Petroleos Mexicanos
Pertamina	Perusahaa Pertambangan Minyak Dan Gas Bumi Negara
PPT	Petroleum Profit Tax
PSC	Production Sharing Contract
Q8	Kuwait refining and distribution company in Europe
R&D	Research and Development
RBI	Resource-based Industrialization
SOE	State Owned Enterprise
SPFI	State Direct Financial Interest (Norway)
Statoil	State Oil company Norway
UAE	United Arab Emirates
WTO	World Trade Organization

CHAPTER 1

INTRODUCTION

The leading theme of this monograph is the role of the state in the economy, and the developing economy in particular, with special emphasis on the role of the state in the oil-producing countries, and the evolution of this role in a dynamic economic environment. This book concentrates on the logic of changing power relations between dominant private and public market players, and the ensuing effect of these changes on the external and internal market organization, in particular the widening gap that private international oil companies and international capital markets drive between the governments of oil-producing countries and their National Oil Companies (NOCs). The pressing need to implement regulatory reforms in the oil sector in the producing countries is demonstrated by the deteriorating performance of the state and their NOCs to run the oil sector efficiently. Furthermore, the international private oil companies' ability to efficiently produce oil in 'higher extraction costs' areas is challenging the NOCs market position. Except for the Middle East, where onshore extraction costs are very low, the extraction cost range has narrowed to $7-$11 a barrel in most of the world oil-producing areas. This means that countries such as Venezuela, Mexico, Nigeria, Algeria, Indonesia, etc., and in particular their NOCs, must compete with private international oil companies. Moreover, oil-producing countries in the Middle East, such as Iran, Kuwait, Saudi Arabia and Iraq, when rehabilitated, are also opening up for foreign direct investments. Access to this 'low-extraction-cost' oil will lower the average cost of a barrel of oil for the international oil companies. At the same time, most NOCs are facing increasing costs. The margin for inefficiency has become much smaller, which requires the NOCs to improve on their past performance.

The fiscal policy of oil-producing countries continues to be very dependent on oil revenues, despite the governments' attempts to diversify the economy. The macro-economic instability of some oil-producing countries, in part caused by international oil market instability, places a premium on stable long-term oil revenues. The NOCs' and OPEC's price and production strategy failed to deliver this policy goal. It is the contention of this book that the state should withdraw from its role as producer and that NOCs should (at least in part) be privatized. In addition, a proper taxation and competition regime should be implemented to ensure that economic rents for the government are secured, the environment is protected, and optimal exploitation of the countries' oil resources is guaranteed.

The point of departure of this analysis is the 1973 oil price increase and the ensuing nationalization of oil assets in OPEC countries. This act of government intervention and the subsequent implementation of OPEC production and pricing policies defined the market structure of the international oil industry for the next decades. The response of the international oil companies to the nationalization of their main production assets was diverse. Some companies withdrew to a particular market and started concentrating on downstream activities. Only a few companies remained focused on the international oil market and used the price policy of OPEC as a cushion to invest in new oil regions. These oil deposits had been uneconomical to exploit before 1973, but became viable investments under the price umbrella of OPEC. The ensuing development of managerial and technological skills and access to capital markets has redefined the competitive conditions on the international oil market. The competitive strategy of the large private international oil companies demands a new approach on the part of the governments of oil-producing countries.

This book has nine chapters. In Chapter 2, the developments on the international oil market since 1973 are discussed in the light of the changing ownership relations in the oil industry. In Chapter 3, the role of the state in the economy is explored. In Chapter 4, the international economic developments since 1973 are considered and the impact these dynamic changes had on oil-producing countries. In Chapter 5, OPEC's failure to secure economic rents for its member states is analyzed, and Chapter 6, evaluates the attempt of OPEC countries to achieve sustainable economic development. In Chapter 7, the relationship between the state and the NOCs is considered, and we explore the competitive position of NOCs in the international oil market. In Chapter 8, an appraisal is given of the consolidation and the competitive reconstruction of the international oil companies after the nationalization in 1973. In the concluding chapter, Chapter 9, we argue that privatization will proceed and will include the largest NOCs, which makes continued inter-governmental regulation of the oil industry, such as OPEC, unlikely. The future of successful privatization of many of the NOCs will be in the hands of the governments in the Gulf, such as the Saudi Arabian government, and their plans with regard to privatization.

CHAPTER 2

REDRAWING THE BOUNDARIES BETWEEN STATE AND COMPANY

2.1 OWNERSHIP

Contrary to a widely held belief, the formation of cartel power on the crude oil market was not the only radical reorganization that took place in the 1970s. Cartelization of the market had occurred in earlier industry growth cycles as well, and is indeed part and parcel of the operation of the oil industry.[1] The novelty for the oil industry was that cartel power was established by producer governments; the novelty for the world was that these were governments of developing countries. These innovations in the power relations in the oil industry had major repercussions on the world economy, and the 'old economic order' was, albeit hastily, proclaimed obsolete. As a result, in the oil industry, it was the state or country that became the dominant level of analysis, while the sector or company level received less attention. This new emphasis rightly focused attention on government regulation of markets, but at the same time resulted in the neglect of the underlying dynamic market forces at company level.

The nationalization of the reserves and production assets of the large private international oil companies generated a large scale de-integration of ownership of the value chain. This led to an important fracture in the structure of the industry, where the incentives for both forward and backward integration were traditionally very strong. Spreading risks and ensuring the security of supply and demand had underpinned the vertical integration and internationalization of oil companies from the early beginnings of the modern oil industry.

Consistent with governments' aversion to the large international oil companies and their ability to manipulate oil income of individual oil-producing countries, the oil reserves and production facilities were brought under state ownership. State ownership of strategic sectors, like the oil industry, was an accepted feature of the modern post-WWII mixed economy. The absence

[1] C. Van der Linde, *Dynamic International Oil Markets*, Oil Market Developments and Structure 1860-1990, Kluwer Academic Publishers, Dordrecht, 1991, pp. 6-44.

of state ownership in this sector prior to 1973 was more exceptional than after 1973, when large-scale nationalization occurred.

Many oil-producing countries changed their constitutions to affirm state ownership over natural resources like oil. Only a few, while asserting state ownership, entered into contractual relationships with private international oil companies, for instance through Production Sharing Contracts (PSCs)[2]. Although, in practice the role of the NOC could be quite dominating in such a contractual context, the continued presence of international oil companies competing for PSCs was a distinctly different situation from allowing only an NOC to operate the oil and gas resources.[3]

The preference for public ownership and operation of natural resources was rather strong in countries in North Africa, the Middle East and Latin America. In their case, the weakly developed state of the private sector in general, the experience of limited contracting opportunities during the concession era, the social and development goals of their governments, and perhaps a strong belief in the benevolence of the state, made nationalization the preferred option.

The rationality of state versus private ownership only came under broad scrutiny in the early 1980s, when states had to acknowledge their failure as efficient producers and had to admit the weaknesses of their monitoring capabilities.[4] It took a little longer to convince the governments of oil-producing countries. The fact that they had regulated the international oil markets through OPEC only since 1973, and very soon found themselves struggling to defend member states' interests in the 1980s, prevented a

[2] In a Production Sharing Contract, the state, usually represented by and through its NOC, takes a more active role in optimizing the development of oil production by forming a partnership between the state or NOC and a private (foreign) oil company. Typically, the foreign company carries out the operational phase of any development. The essential elements in a PSC are cost recovery by the foreign oil company through 'cost oil', (i.e., the company takes oil to recover its costs); production sharing (after the deduction of 'cost oil' the remaining oil is called 'profit oil' which is shared according to an agreed key with the state) and tax (payable by the foreign company on its share of the 'profit oil'). D. Fee, *Petroleum Exploration Strategy*, Belhaven Press, London/New York, 1988, pp. 34-35.

[3] P. Barnes, *Indonesia, The Political Economy of Energy*, Oxford University Press, Oxford, 1995, pp.142-58

[4] A. Shleifer, 'State versus Private Ownership', *Journal of Economic Perspectives*, Fall 1998, vol. 12, no. 4, pp. 133-50.

quick turn around on nationalization. Their efforts were geared towards getting the OPEC policies right rather than getting national policies right. One might even go as far as to say that OPEC, as it functioned in the 1970s and 1980s, may not have combined well with private ownership, because it would have diminished the decision-making abilities of the oil minister at the OPEC conference. However, towards the end of the 1980s, the mood began to change, and an admission that state oil companies were not so efficient may be deduced from a number of attempts that were made to commercialize the companies. Furthermore, the privatization in the former Soviet Union placed additional pressure on OPEC countries. Yet, the reluctance in OPEC countries to reverse the nationalization was very great, and, instead, it was non-OPEC countries like Argentina that took the lead.

The new insight into state versus private ownership is quite straightforward in its recommendations, and, if it is adopted, will have large implications for state oil companies. The case for government ownership in a country with good contract enforcement and a benevolent government is rather limited. Circumstances that would make government ownership preferable are, according to Shleifer [5], situations in which: "(1) opportunities for cost reductions that lead to non-contractible deterioration of quality are significant; (2) innovation is relatively unimportant; (3) competition is weak and consumer choice is ineffective; and (4) reputational mechanisms are also weak." It may be said that public ownership provides cost savings when cost reductions and innovation are not essential, and when the incomplete contracting is caused by the ambivalence of the government with regard to what to produce. If the government does not want to pay if they change their mind, public ownership has benefits. Those familiar with the oil sector suggest that this description reflects the experience of state-owned oil companies in certain countries up to the early 1970s, by which time patterns in the industry were beginning to change, leading in turn to the need for a new set of monitoring and ownership policies. Yet, even if the policy consensus about the role of the state in the 1970s had been different, the absence of a sizeable private sector in these oil-producing countries, once the market power had shifted, would have prevented any other thoughts than nationalization.

The case for private ownership becomes even stronger if we take the imperfect state as a point of departure and political goals become part of the

[5] *Ibid.*

equation. In order to maintain political support, governments have a strong incentive to distribute benefits to their supporters, either in direct transfers, in jobs or in jobs at an above-market wage. Shleifer observes that "Governments have used their control of state firms and other assets as a means of channeling these benefits, by forcing excess employment at state firms and agencies, creating government projects that transfer wealth to supporters, and so on. In other words, state firms are inefficient not just because their managers have weak incentives to reduce costs, but because inefficiency is the result of government's deliberate policy to transfer resources to supporters."[6] Needless to say, the gains from privatization are much greater when an imperfect state is involved.

The decision of the oil-producing countries in the 1970s to nationalize the oil industry in their countries was not different from the practice of other developed countries, and was accepted as a natural outcome of the changing relations between oil companies and host countries. In the United Kingdom and Norway, state oil companies were given a role in developing the North Sea oil assets.[7] The dominant role of government in the oil sector in OPEC countries was by no means exceptional.

Nevertheless, the nationalization created considerable anxiety in consumer countries with regard to security of supply, particularly when the Arab oil-producing countries felt few inhibitions about using their oil as a weapon to change the outcome of the October War in 1973.[8] Their sudden awareness of their great strategic dependence on oil imports caused governments in consumer countries to shift their energy policies away from oil, or at least imported oil.[9] These revised policies included the implementation

[6] *Ibid.*

[7] Mabro et al., *The Economics of North Sea Oil*, Oxford, OIES/OUP, 1986. In the UK, BNOC (British National Oil Corporation; established in 1975) produced oil until 1982 (share of UK Continental Shelf was 31.8% and 24.5% of total North Sea production), when Britoil was privatized and BNOC continued as trading company until its abolition in the mid-1980s. In Norway, Statoil operates the Norwegian interest in the oil sector (its share was 37.2% of Norwegian production and 8.3% of total North Sea production). The shares in total North Sea production of the major international oil companies (BP, Shell, Exxon) were 13.5%, 11.4% and 11.2%, respectively. See also J. Vickers and G. Yarrow, *Privatization: An Economic Analysis*, MIT Press, Cambridge, Mass., 1988, pp. 316-22.

[8] A. Sampson, *The Seven Sisters*, Coronet Books, Sevenoaks, 1975, pp. 262-72.

[9] C. van der Linde, *op. cit.*, pp. 99-112.

of investment incentives to stimulate the development of oil production in 'safe' countries. The oil price increase had made investments in previously uneconomic oil assets, such as the North Sea, much more feasible. These investment incentives were, however, not only directed at the large international oil companies, despite the fact that they were about to surrender control over the bulk of the oil reserves and production assets, but also to stimulate newcomers in the market. The recognition of the consumer countries of their dependence on only a few exporting countries included an awareness of their dependence on only a few international oil companies.

The position of the large international oil companies was fragile. There was little sympathy with the international oil companies, particularly when it became clear that, while consumer country economies were struggling with the impact of the oil price increase of 1973-1974, oil companies were recording tremendous windfall profits. In their home markets, governments stimulated the entry of new companies by auctioning off smaller concession blocks and awarding some of these exploration contracts to newcomers or smaller independent companies.

Whereas the 1970s and early 1980s were the years in which oil produced for the international oil market was progressively brought under state control, the late 1980s and 1990s showed a contrary development. The oil price fall in 1986 and the continued lower oil prices in the years thereafter unleashed a move away from the, indebted, state. In the 1980s and early 1990s, critical discussions on the rationality of state ownership began to gain ground among policymakers both at national and international level[10]. Although the oil industry was still slow to adopt these insights, just as it was late in bringing it under state control, some countries, in developed as well as in developing countries, took the bold step of privatization. In the UK, Britoil, the state oil company was stripped of its special status, and the government also sold its large block of shares in BP. The liberalization of the oil industry in developing countries was, admittedly, largely 'inspired' by International Monetary Fund (IMF) stabilization programmes. Argentina and Brazil fit in this category. Other countries, not under such international pressure, did see the need to at least commercialize state oil companies, and in the 1980s we saw some attempts to make the state oil company more like

[10] D. Helm and T. Jenkinson, 'Introducing Competition into Regulated Industries'. In D. Helm and T. Jenkinson (eds.), *Competition in Regulated Industries*, Oxford University Press, Oxford, 1998, pp. 1-2.

a 'normal' company. In this from the point of view of the NOCs more competitive market, in which government control has become less direct, state oil companies are trying to redefine their roles. This redefinition of state company strategies has put them on occasion at loggerheads with their governments. PDVSA, the state oil company of Venezuela, developed its own company strategy in the late 1980s, but the implementation was restricted by the state.

A major boost in the privatization of state oil companies in the oil industry really came in the early 1990s, when centrally planned economies started on a process of transition towards market economies.[11] The mounting problems in the Soviet oil industry in the 1980s[12] led to drastic reforms of oil market regulation and control.[13] The fragmentation of the Soviet Union into separate states and their painful transition to a market-based economy opened up the oil sector for private investment.[14] The liberalization of the oil industry in the former Soviet Union has changed the competitive position of all oil-producing countries. Private capital was in high demand, and earlier inhibitions in allowing foreign investors to own resources were reversed.

The availability of many investment opportunities in the 1990s has intensified the competition for scarce capital and technology by the oil-producing countries. The international private oil companies, after a period of nearly 20 years, are again playing a dominant role in the international oil industry. Although they are less often seeking 100% equity stakes than in earlier periods, and are more aware of investment risks, their control over the market has increased. Joint ventures, strategic alliances and, of late, mergers, are serving to balance risk and control for the large private international oil companies thus putting them in a much better position to maintain profitability than the state oil companies, despite lower oil prices.

[11] See generally J.E. Stiglitz, *Whither Socialism*, MIT Press, Cambridge, Mass, 1995, pp. 171-96.

[12] See generally C. van der Linde, 'Empty Barrels Make the Most Sound', *Journal of World Competition*, vol. 15, no. 2, Geneva, 1991; T. Gustafson, *Crisis amid Plenty*, Princeton University Press, 1989.

[13] See generally M. Messer, *The Reorganization of the Russian Oil Industry*, 1990-1997, draft Doctoral Thesis, Leiden, to appear 2000.

[14] J.E. Hartshorn, *Oil Trade: Politics and Prospects*, Cambridge University Press, Cambridge, 1993, p. 276.

2.2 DEVELOPMENT STRATEGY

The massive entry of governments into the market, not only as shareholders in their NOCs but also as participants in a cartel, brought political concerns and economic welfare much more to the fore in the industry. Governments' oil policies had to serve wider economic goals than the long-term maximization of profits and had to ensure control of the private international companies they employed. Governments pursued their goals at international, national and sectoral level, and their policies were intended to support the government's overall goals for growth and development. In the 1970s, the majority of oil-producing nations were developing countries with industrialized economies that had relatively little diversification compared to more developed economies. They intended to use their oil market power, reflected in their massively increased economic rents from oil in the early 1970s, to generate a 'leap forward' in industrialization and the modernization of their economies. The idea was that the oil industry would generate investment capital for a less oil-dependent international economy or for a post-oil future of the country itself. The governments saw the NOCs as important vehicles with which to bring about this 'industrial revolution'.

The oil price increase in 1973 forced the oil-producing countries to address the problem of how to invest their windfall profits. In general, the windfall lifted the earlier constraints on economic growth, namely foreign exchange, savings and fiscal revenues. Investing the profits had to generate income in the long term. The oil-producing countries' pattern of investment reflected the following objectives: the expansion of growth of the non-oil economy, the modernization and diversification of the economy and reduced dependence on oil, and consolidation of national control over the oil, gas and other industrial sectors.[15] The bulk of their investments (about two-thirds) were directed towards to infrastructure or human capital formation and therefore supported the non-traded sectors. The remaining investments were put into energy or energy-intense industries.[16] Although in this way they appeared to acquire a comparative advantage in these traded sectors, they did not always manage to capture these advantages, and the riskiness of these types of investments was not always properly taken into

[15] A. Gelb and Associates, *Oil Windfalls, Blessing or Curse*?, A *World Bank* Research Publication, Oxford University Press, Oxford, 1988, p. 11.

[16] *Ibid.*, p. 137.

account. When resource-based industries slumped in the 1980s, the oil-producing countries encountered grave economic problems. Apart from a few exceptions, their dependence on oil had not diminished, and in some cases had even increased. Their fiscal dependence on oil income remained great, and fluctuations in this income defined the space, or lack thereof, governments had to take particular policy decisions. The sheer size of their windfall profits in 1973-1974 and in 1979-1980 had made the countries vulnerable to the 'resource curse' or 'Dutch Disease'[17].

The economies of the oil-producing countries did not fare well. After the 1986 price collapse, countries such as Mexico, Venezuela, Nigeria and Algeria encountered severe economic and debt problems that required the assistance of the IMF. Mexico encountered new problems in 1994-1995, and negatively influenced other economies in Latin America, including smaller oil producers, like Ecuador, Argentina, and Colombia. In Africa, both Algeria and Nigeria have experienced catastrophic political problems, which have caused their economies to slip even further. In the Middle East, Iran and Iraq were engaged in a war of attrition in the 1980s, from which Iran is only slowly recovering. Iraq's economy, after the long war with Iran, the war in Kuwait in 1990-1991 and the ensuing UN economic embargo, is almost non-existent. Indonesia's economy was severely damaged in the Asian crisis, and the country had to seek assistance from the IMF to stabilize the economy and restructure its debts.[18] The Russian economy has yet to emerge from its process of transition to a market economy, and remains under IMF scrutiny. The financial crises in Asia in 1997, in Russia in 1998 and in Brazil in 1999, combined with low oil prices, have yet again exposed the weakness of the economies of the oil-producing countries, and their considerable dependence on the oil sector. In the longer run, the benefits of the windfall profits in the 1970s and early 1980s have turned out to be very small.

2.3 CONFLICTING POLICIES

Over time as governments pursued their goals at the above-mentioned three different levels, they increasingly found that these goals conflicted with

[17] S. Van Wijnbergen, ''The Dutch Disease': A Disease after all?', *Economic Journal 94*, March, 1984, pp. 41-55; A. Gelb et al., *op. cit.*, 1988, pp. 14-45.

[18] C. van der Linde, 'Hervorming van het IMF', *Internationale Spectator*, jaargang 52, no. 11, November 1998, pp. 576-81.

each other. The governments coordinated price and production policies within the OPEC framework to collectively secure higher price levels and to optimize income from oil in the international market. At the same time, national oil policy had both to secure optimal present and future income from oil to facilitate a diversified industrialization strategy, and to generate general economic development. The NOCs were thus established as an instrument of government policy to ensure sovereignty over the nation's resources and to distribute the welfare derived from these national resources.[19]

The OPEC countries' three-pronged strategy did not take place in a vacuum. International private oil companies and consumer countries were strongly affected by the implementation of OPEC policies in the 1970s and early 1980s. The two oil price increases brought about new supply and demand relations. The governments of consumer countries, especially industrialized countries, implemented policies to increase energy saving and efficiency, and diversified their dependence on imported energy and oil. These policies changed the demand and supply conditions in the market, which had, in turn, to be taken into account by the governments of the oil-producing countries and the NOCs in order to secure their interests. Furthermore, the price increase of 1973 turned out to be the pinnacle of OPEC's success. In game-theoretic terms, OPEC member states had enough interests in common in 1973 to agree and reach a Nash equilibrium.[20] From mid-1974 onwards, however, the asymmetry in interests as a result of different economic determinants undermined the member states' ability to make credible commitments to cooperate within OPEC. Progressively, OPEC's weakness became apparent.

The major distributional disputes that occurred among OPEC member states in the 1980s, and that had rapidly reduced the organization's market power when the oil market turned into a buyer's market were repeated within the member states themselves as escalating sectoral distribution tensions. The newly-formed NOCs had been assigned the task of generating income from oil at company level. Governments secured the economic rents and profits from the state companies through either taxation or direct claims on a large percentage of pre-tax income from the NOCs, and redistributed

[19] M.A. Olorunfemi, 'The Dynamics of National Oil Companies', *OPEC Review*, Winter 1991, volume xv, no. 4, p. 322.

[20] C. van der Linde, *op. cit.*, 1991, p. 26.

the proceeds through the economy. Obviously, the state companies, acting on their governments' wider economic (and sometimes political) interests, could not easily develop efficient company strategies, since production levels were at the discretion of the oil minister, and, the companies were also frequently forced to sell oil products below cost on the domestic markets. As long as the OPEC countries were able to generate enough income from oil to enable them not only to increase investment in the oil industry but also to redistribute the profits and economic rents to other sectors of the economy, the strategy of the government seemed successful. However, when income from oil decreased in the early 1980s, and production levels had to be curbed, the ambiguity of the governments' policy became clear to the NOCs. Their ability to expand into mature and integrated oil companies on the world market became hampered when increasingly investment plans were shelved or canceled. Most oil-producing countries, like Mexico, Venezuela, Algeria and Nigeria, encountered economic and financial problems in the 1980s and 1990s, when their national expenditure outpaced national income. The pressure on the NOCs to generate more foreign currency income was great, while at the same time governments continued to force these companies to sell oil products in the domestic economy at below production cost.

2.4 National Oil Companies

The NOCs vary widely across the oil-producing countries. In some countries they exercise complete state ownership in exploitation and in others they collaborate with other companies in a licensing regime.[21] Production Sharing Contracts (PSCs) represent an intermediate form. In some countries (e.g., Nigeria), the NOC is the joint venture partner of international oil companies, representing the state's interest in each activity. The NOC's share in the partnership is usually around 50%, and sometimes less. In Indonesia and Malaysia, the NOC enters into PSCs, in which the international oil company is the operator of the activity and supplies the capital

[21] R.W. Bentham and W.G.R. Smith, *State Petroleum Corporations: Corporate Forms, Powers and Control*, The Centre for Petroleum and Mineral Law Studies, University of Dundee, 1986, pp. 3-4.

and technology.[22] In other countries, the NOC operates the oil exploration and production facilities, oil processing and transport. Kuwait (KOC/Q8) and Saudi Arabia (Saudi Aramco) also have interests in some market refineries in the US and Europe. In Saudi Arabia, the petrochemical sector is operated by SABIC, another state-owned company, in partnership with international oil companies. Recently, joint ventures have become an important form of alliance between NOCs and international oil companies. The joint venture arrangement can support the NOC, which requires additional investment capital and contract expertise to develop new fields that it is prevented from doing itself. This is a significant departure from the period of nationalization, and shows that even the staunchest defenders of a nationalized oil industry are shifting their position.

In the case of Venezuela, PDVSA was a proponent of bringing international investors in, while the government took a long time to be convinced of the merits of such a move. The government's reluctance to allow private foreign investors into the oil industry can be partly explained by political motives. For many years, the government had struggled to increase its authority over the powerful private international oil companies, nationalization finally being achieved in the 1970s. Allowing private foreign investors back into the oil sector, despite the changed circumstances in contractual relationships and the possibilities for regulating the sector, could be seen as reversing the politically much heralded establishment of national authority over natural resources.

The degree of independence at managerial level also varies widely among oil-producing countries. In most of them, a close relationship exists between government officials (ministers of oil, mines and finance) and top management of the NOCs.[23] Job rotation from the NOC to the government and sometimes back is quite common. This close relationship does not allow for much independence on the part of NOCs' management. Firstly, a job in the government is part of a successful career in the NOC, and secondly, government ministers coming from the NOC might find it hard not to interfere with day-to-day management of the company. Moreover, too much sympathy with the arguments of the NOC's senior management

[22] T.N. Machmud, *The Production Sharing Contract in the Indonesian Petroleum Industry: At a crossroads after 30 years of history...... status quo or reform?*, draft doctoral thesis, Leiden University, Chapters 3 and 4, to appear 2000.

[23] R.W. Bentham and W.G.R. Smith, *op. cit.*, pp. 10-96.

might put the minister at loggerheads with his fellow-ministers in the government. This complicated interpersonal relationship between government and NOC may muddle the waters when interdepartmental interest-seeking is involved, particularly when the ministry of finance (or as in Indonesia the Central Bank) handles the finances of the NOC. The few attempts to distance the NOC from government without any formal arrangement in countries like Venezuela and Mexico[24] have so far been unsuccessful. This is partly because there is too much at stake for the government, and partly because the government and the NOCs disagree as to whose interests should prevail, the NOCs' or the governments'. The NOCs are very important to the national economy, and the government does not want to lose control over such an important part of the economy. Moreover, some NOCs are so large compared to the domestic economy that the NOC might become a state within a state. For their part, the NOCs are operating in an international competitive market and wish to have more freedom to capitalize fully on their competitive advantages, which have been diminished due to the organizational failures associated with their form of ownership.

NOCs, especially during the years of economic hardship from the 1980s onwards, were used as rent-collectors, and were left little room to develop independent business activities. In addition to the overt task of operating the oil industry for the government, the hidden task of the NOCs was to be a vehicle for economic development. This dual role as a development agency imposed certain limitations on their business strategies, as they were supposed to invest in non-sector-specific operations, boost employment in the country, and subsidize domestic demand for oil products. The fact that NOCs were an extension of government policy was also a hindrance to their internationalization strategies. The investment of Kuwait in British Petroleum (BP) was reversed on legal grounds by the Monopoly and Mergers Commission of the British government. The strategic political and economic interests of the UK were deemed to be at stake. Moreover, NOCs were also hindered in their internationalization strategies by their own governments. Investments in assets abroad could, if financed through the inter-

[24] In the present election campaign, R. Madrazo, a main contender for the PRI's presidential nomination, proposed to issue shares of Pemex on the stock market and open itself up for private foreign investment. *Financial Times*, 17 August 1999.

national capital markets, have a negative effect on the country's debt ratio, and could therefore be prevented by government decision.

Increasingly, the credit rating of NOCs unsurprisingly outpaced that of their governments, putting more pressure on the companies to generate funds for the cash-starved economy. The NOCs were competing in an international market, and increasingly attempted to secure demand for their oil through downstream investments in the main consumer markets rather than expanding exporting facilities at home. This strategy, at times, created tension between government officials, who wanted domestic investment and employment, and the management of the companies who wanted to pursue an international business strategy.

The present initiatives in certain OPEC countries to modify the domestic oil market organization and the relationship between the state and the NOC must be seen as a logical extension of the ongoing changes on the international oil market. Economic institutions, government departments and companies alike cannot extricate themselves from the market process, however strong their market power at a given point in time may have been. Many studies of the emergence and operation of OPEC's cartel power in the 1970s and 1980s left readers with the impression that the structure of power in the oil market would last in perpetuity, as had been the belief in the decades before that the 'Seven Sisters'[25] would always dominate the market. The power to regulate the market, it was argued, derived from ownership of actual and potential production capacity by the dominant producers, who, in turn, generated power over the pricing of crude oil. In both cases, these conclusions were underpinned by reserve and production statistics. Nevertheless, despite the statistics on potential production capacity, OPEC's market power has disappeared and the prolonged weakness of the crude oil market has forced its individual member states to focus on the reorganization of their national industry. Moreover, many of the member states can no longer generate internally the capital required to finance the next generation

[25] The 'Seven Sisters' were the seven oil companies that dominated the international market from 1945 until 1973. The group consisted of: Exxon, Chevron, Mobil, Texaco, Gulf, BP and Shell. Through mergers and takeovers (Chevron/Gulf and Exxon/Mobil) the original group has been reduced to five. The difference in size among this group was substantial. In the 1980s Exxon, Shell and BP were considered first tier companies, while the others belonged to the second tier. The present merger and takeover wave seems like an attempt for leadership among the first tier companies. See generally A. Sampson, *op. cit.*, 1988.

of investments in production capacity and/or processing plants. As a result, they are increasingly courting private international oil companies. At whatever level they will eventually readmit the private international oil companies to their oil industries, it will strongly impact national oil policies, the role of the state and the position of the NOC.

2.5 International Oil Companies

The nationalization of private oil companies' reserves and production assets in the major oil-producing countries had greatly upset the major oil companies' balance between upstream and downstream activities. This forced process of backward de-integration radically changed the market position of the companies, with their main assets now in processing, rather than in production, and geographically less dispersed. In terms of their core skills, the large international oil companies lost important site-specific and dedicated assets, and found the boundaries of their business involuntarily redefined. Although the international oil companies immediately stepped up exploration and production in 'investment-safe' areas like the North Sea and Alaska, these investments could not effectively replace their former vast rent- and profit-generating activities in OPEC countries. Rather, the production of higher-cost oil compelled the oil companies to seek alternative ways to generate rents and profits, and to regain their competitive advantage through R&D and innovation.

The nationalization of a substantial part of the upstream assets of the private oil companies meant that they needed a new internal organization, as well as a new market strategy more in line with their new strategies and company activities. The nationalization of the oil assets in OPEC countries had shifted them predominantly into the downstream part of the value chain, and their geographic scope had been temporarily narrowed.

The refining industry was particularly affected by the price increases in crude in 1973-74 and 1979-80. Furthermore, the refineries in the consumer markets were equipped to process predominantly lighter crudes, which gave them the best yield on the refined product mix. Particularly in Europe and Japan, fuel oil was used to generate electricity. The decision by governments in consumer countries to move electricity generation away from fuel oil caused the chemistry of demand to move towards the lighter end of the barrel, while supply remained at the heavier end of the barrel. The mismatch between refinery input and output was exacerbated by the fact that crude supply was becoming heavier in the 1970s. This confronted the international oil companies with another imbalance in their operations. The large international oil companies resolutely restructured their refining sector. In addition to closing excess capacity in the more mature markets, they made their remaining refineries more flexible to handle different types of crude (i.e., crude switching) They also upgraded into more complex refinery techniques to produce lighter oil products from heavier oil. After substantially reducing their refining capacity in the mature markets and upgrading their remaining capacity, they built new refining capacity in Asia, where demand for refined products was strong.[26] In contrast, the smaller, non-integrated refineries in the US did not restructure on the same scale as the large integrated oil companies. The small refineries were supported by US government policies until the 1980s, after which time also the small US refineries industry reorganized. However, this process of rationalization was slow, and only sped up when environmental legislation required substantive upgrading. As a result, many small refineries were closed. In general, twenty years of rationalization and upgrading has increased concentration in the US refining sector.[27]

As a result of the disintegration of their value chain, the international oil companies had to innovate in the area of their core skills, develop new competitive advantages and define new boundaries of the firm. This sort of development has also occurred in other industrial sectors[28]. Raw material production (oil) was taken over by developing countries' local producers (or state companies) and some simple processing industries. The large international companies (MNCs) subsequently move upwards in the value chain, while the production stage and bulk processing is carried out by smaller companies or state companies in the producing country. The strategy of the international companies to secure and control supply of the raw material (oil) as before (i.e., through vertical integration), entailed the development of frontier technologies in the exploration phase and improved efficiency of production at all stages of the value chain. In addition, new products

[26] C. van der Linde, *op. cit.*, pp. 193-5.

[27] K. Hamm, 'The Refining Industry in North America', *The Energy Journal*, vol. 15, Special Issue, 1994, p. 183. See also K.R. Baliker and J.W. Forest, 'Stretching To Reformulate', *Petroleum Economist*, April 1992, p. 12.

[28] W. de Haan, F. van der Molen and C. van der Linde, 'Een Zwaargewicht slankt af: Unilever 1965-1980'. In F. Crone and H. Overbeek, *Nederlands Kapitaal over de grenzen*, SUA, Amsterdam, 1981, pp. 123-67 .

were developed at the higher end of the value chain, and gas production, transportation and distribution became part of the international oil companies' core business.

By developing superior exploration and management skills, they would be able to create a competitive edge in non-OPEC oil exploration, which was predominantly offshore or in hostile environments, and reduce the minimum required crude oil price at which they could compete in the international oil market. Bringing the cost down of frontier oil-field production would create a competitive fringe for OPEC crude oils, effectively undermining the cartel power of OPEC. Moreover, this upper cap on prices would protect final consumer markets, where the international oil companies were still very strong.

The international oil companies had to use the window of opportunity that OPEC provided in the early 1980s with the stabilization of oil prices at a fairly high level, and they had to reduce costs before prices declined again. This strategy was not without risk. The vulnerability of the international oil companies was two-fold. At high crude oil prices, their processing margins would be slim and their final markets would shrink; while at very low prices, their frontier investments would turn out to be unproductive.

The key element in this strategy is the imposition by the international oil companies of a new phase on the value chain, a phase preceding exploration: exploration R&D. Although the initial costs of developing new exploration and production techniques were high, the international oil companies could develop these techniques under the price umbrella of OPEC and recoup their investments. When these techniques were mastered, costs began to decline, and oil produced in these initially high cost areas became cheaper. The price band in which the oil companies could profitably find, develop and produce non-OPEC oil, and in particular oil from frontier oil fields, came down substantially (in terms of 1999 prices from $25 a barrel in the 1980s to $10 in 1999[29]). The volume of non-OPEC oil on the market increased rapidly in the period 1980-1986, when OPEC insisted on a policy of restricted production to support the price. Although the competitiveness of low-cost oil from the Middle East remained unchallenged, these OPEC policies resulted in oil from higher-cost fields was increasingly replacing oil produced in low-cost fields. In 1985-1986,

[29] *The Economist*, March 6, 1999, p. 22.

Saudi Arabia led OPEC into a lower price regime/higher market share strategy, and low-cost oil recouped its market share. The lower oil prices did stunt the growth of exploration in frontier fields, but the operating fields continued to produce, albeit at lower or zero profitability rates. It was pointed out that "even in the North Sea, the bare-bottom operating costs have fallen to $4 a barrel. For the lifetime of such fields firms will continue to crank out oil, even though they are not recouping the sunk costs of exploration and financing."[30] Only when oil prices sink below a level at which operating costs can no longer be recouped are oil fields taken out of production. That is, private companies will take these fields out of production. Countries that are extremely dependent on oil for government income and hard currency may decide to continue producing oil for some time, whatever the price.

The learning curve of the large international oil companies that embarked on frontier oil exploration in the 1970s and 1980s has been steep. The technological innovations in finding, developing, and producing oil have been significant. The level of technological and managerial skills is very high now and has increased the barriers-to-entry in higher cost oil exploration for companies that did not engage in this type of investment. The NOCs in particular have fallen behind. Since the 1970s, the NOCs have been brimming with cheap onshore oil reserves and, moreover, have been restrained in their oil-producing in order to support the oil price. They did not need to invest in new exploration technologies until their reserves needed to be replaced. Governments preferred to spend their oil income elsewhere in the economy, rather than allowing the NOCs to engage in R&D activities. In the longer run, this has harmed the ability of the NOCs to compete. New explorations in many oil-producing countries are taking place in the offshore rather than the onshore fields, and the NOCs have not sufficiently developed their technological capability to bring this oil competitively on stream. In order to incorporate the required technical changes into existing production systems, the NOCs will have to draw on external suppliers for engineering services but may have to generate a significant part of the technology themselves if they wish to continue playing an independent and active role.

The international oil companies' efforts in exploration gave them substantial leverage over the market because they could progressively supply

[30] *The Economist*, March 6, 1999, p. 22.

the market, and more particularly their refineries, with lower-cost marginal oil. Lowering the price of the marginal barrel of non-OPEC oil meant that more oil came onto the market that was able to compete at the fringe with the low-cost oil from OPEC. Although compared to non-OPEC oil, OPEC oil was very cheap to produce, it is important to understand that already before 1973 production costs in the OPEC countries were not homogeneous, but varied between the extremely low-cost oil in the Gulf region and the relatively higher-cost oil in Venezuela.[31] Over time, the cost of continuing production from these oil wells also entails costs when secondary and tertiary production methods have to be introduced to maintain production. The cost of bringing new oil deposits into production also increased in the OPEC countries, as the new developments moved increasingly off-shore.

Macro-economic instability, and particularly inflation, can increase the cost of investments for state oil companies, particularly when the investments need to be financed on the international capital markets. The relatively low price of oil over the past 15 years have reduced the value of the companies' collateral and, in some cases, large public debts imply higher premiums on interest rates.

The competitiveness of OPEC oil also declined as a result of inefficiencies on the part of operating state oil companies. As noted earlier, the political objectives of government bureaucrats in OPEC countries caused problems such as overstaffing, payment of above-market wages, and other transfer problems in the state oil companies. Overhead costs must be assumed to be fairly high, while at the same time management has little incentive to innovate and increase efficiency.

The improvement in the competitiveness of the oil produced by international oil companies in conjunction with a reduction over time in the competitiveness of OPEC oil made it harder for OPEC to increase prices beyond a certain level unless its member states were willing to suffer large losses in market share and/or income from oil. The economic situation of most oil-producing countries in the 1980s made this a progressively unlikely option. The international oil companies' strategy of lowering the production cost of the marginal barrel of non-OPEC oil, combined with

[31] M. Adelman, 'Crude Oil Production Costs (1966)' In *The Economics of Petroleum Supply*, 1993, MIT Press, Cambridge Mass., p. 61.

upgrading the downstream process and final products was, under the OPEC regime, a rational strategy.

This strategy did not, however, fully protect the oil companies from the peril of prices falling below the price band. The economic deprivation of some oil-producing countries pressured them into producing more oil to satisfy their short-term need for foreign exchange. Moreover, the revival of the international oil industry in non-OPEC countries created excess capacity in the international oil market.

A criticism often made of the oil industry is "that it is not self-adjusting".[32] This lack of self-adjustment of the oil industry is due to: "(1) the aleatory character of drilling coupled with the high costs of exploration and the low cost of exploitation; (2) the unwieldy relation of fixed and variable cost in refining, transport, and marketing; and finally (3) a price structure that allows for ups and downs which fail to bring relief from dearth and glut."[33] The major factor that appears to introduce adjustment is regulatory intervention by a market leader or cartel. For example, Frankel argued that the cost of price stabilization would have to be shouldered by the market leader or cartel, while the negative aspects of such a concentrated market should be countered by actual or potential entry into the market by smaller companies. The implementation of anti-trust legislation with varying precision gave government an important say in the organization of the national oil industry.[34]

Adelman agrees with Frankel's view of the unwieldy relationship between exploration and exploitation cost.[35] However, he is less negative about the self-adjusting abilities of the international oil industry in the

[32] P. Frankel, 'Essentials of Petroleum: A Key to Oil Economics' (1946). In I. Skeet (ed.), *Paul Frankel Common Carrier of Common Sense, a selection of his writings 1946-1988*, Oxford: OIES, 1989, p. 27.

[33] *Ibid.*

[34] The relaxed view of the anti-trust legislators with regard to upstream concentration (BP Amoco's takeover of Arco resulted in concentration of Alaskan oil production) and the precise implementation of this legislation in the refining and distribution sector (Arco's operation on the US west coast is very efficient, where BP Amoco is weaker), is a case in point of the degree of regulatory freedom of governments in regulating competition, (c.f. *The Financial Times*, 30 March 1999).

[35] M. Adelman, 'Efficiency Of Resource Use in Crude Petroleum (1964)'. In *The Economics of Petroleum Supply*, MIT Press, Cambridge, Mass., 1993, pp. 28-30.

longer term. Low oil prices do not lead to the immediate closure of production facilities as long as operational costs are covered, but low oil prices will reduce investments in new production facilities. Thus, in the longer term capacity will decline and the industry will recover.

Jensen recognizes a general tendency to delay exit. He observes that "managers fail to recognize that they themselves must downsize; instead they leave the exit to other while they continue to invest. When all managers behave this way, exit is significantly delayed at substantial cost of real resources to society."[36] The problem of exit is particularly large in the oil industry. Low prices do not sufficiently clear the market of excess capacity, because producers will continue to produce oil as long as they can cover the operating costs.

In the case of state oil companies, exit is not an issue at all. Given the constitution of the present international oil market with private international oil companies and state oil companies, excess capacity cannot easily be removed from the market. The decline of OPEC as a market regulator, the private-public ownership of oil assets, and the general economic downturn in oil-producing countries caused production to overshoot demand and oil prices to fall to very low levels in the 1990s.

The international oil companies' strategy, which was conceived when oil prices were fairly high and when ownership of cheap oil reserves was solidly in the hands of the oil-producing countries and their NOCs, could not prevent them from getting hurt in the oil price plunge in the 1990s. When in 1998 oil prices declined to very low levels ($10-$12 a barrel), they actually sank under their price band. Since expectations were that oil prices would remain low for some time, the international oil companies decided to engage in a major cost-cutting operation. New investments would only receive the go-ahead if they could be profitable with oil prices of about $10 a barrel. Other investments were shelved.

Some of the cost-efficiencies in the oil companies were pursued by merging with or taking over other companies, with whom they claimed to have such synergies that cost reductions could be achieved. In particular, the BP Amoco and later also the Arco merger/takeover was justified in this context. The fact that this merger/takeover will allow BP Amoco to control

[36] M.C. Jensen, 'The Modern Industrial Revolution, Exit, and the Failure of Internal Control Systems', *The Journal of Finance*, vol. XLVIII, no. 3, July 1993, p. 847.

a large share of Alaskan oil production and can now operate these fields from a single organization should reduce costs. Moreover, BP did not have a strong American-based processing and distribution network. After it bought the remainder of the shares in Sohio to create BP America, its base was still too narrow. Amoco added to its activities in the US, but also to the international portfolio. Arco has a strong US West Coast base, and has significant activities in Indonesia.[37] The synergy benefits seem present feasible, although mergers and acquisitions are usually quite costly when it comes to merging the different company cultures.[38] Moreover, some activities will have to be discontinued, either because there is internal company overcapacity or because the competition authorities in certain markets/countries make this requirement.

The recent merger of Exxon and Mobil, although cost reduction elements and synergies also play a part here, emphasizes another major justification for merger and acquisition activities. Mobil had indicated that it was looking for a 'friendly takeover', because it deemed itself unfit to compete in international oil markets without a major restructuring process. Mobil had substantial downstream activities in the US, but after the nationalization of Aramco, it was crude-poor and not very internationally diversified. Despite its important holding in the Caspian Sea region, Mobil could not really compete fully with the largest oil companies in all markets, existing or new. Together with Chevron and Texaco, the company belonged to the second tier of large international oil companies. As a result of the low oil prices in 1998 and the aftermath of the Asian Financial Crisis, oil share values, including Mobil's, were fairly low. Rather than wait for an unfriendly takeover, it put itself on the market. Exxon, in addition to its interests in synergies both in the US market and in the Caspian Sea region, could or would not allow a company like Mobil to go to a competitor. The merger will also bring about disinvestment because before the merger will be effectuated, the competition authorities in the US and Europe might instruct the company to discontinue part of the joint activities due to dominant downstream market positions.

The international oil companies developed new core skills, and , as long as oil prices remained within the $12-18 price band, their technological and

[37] *Financial Times*, 30 March 1999.

[38] H. Schenk, 'Policy Implications of Purely Strategic Mergers', *Erasmus University Management Report*, no. 21-1999, p. 11.

managerial skills enabled them to compete with 'low- cost' oil, produced by less efficient NOCs. The prospect of new ownership relationships in oil-producing countries (Russia, Kazakhstan, Indonesia, etc.), combined with the expectation of a fairly long period of oil prices in the region of \$10-\$12 a barrel, compelled them to re-engineer their cost and organization structure in order to continue to capture a large part of the information and human capital rents.

2.6 CONCLUSION

The international oil market is in a flux. Ownership of oil reserves and production facilities has again appeared on the agenda. This time the state has begun a process of withdrawal from its role as a producer. This withdrawal can be partly explained by the prolonged economic problems of many oil-producing countries, and the awareness that other systems of ownership regulation might optimize their income from oil. The international oil companies play an important part in this changing structure of the international oil market. They will have to provide the capital and technology. However, the transformation to a new organizational structure of the international oil market is so slow that low oil prices are undercutting their performance as much as the performance of state oil companies. In 1998 and 1999, this already prompted substantial realignments of company strategies and led to increased merger and takeover activity.

CHAPTER 3

STATE PARTICIPATION IN THE ECONOMY

3.1 INTRODUCTION

Government intervention in the economy has been decisive for the development of oil-producing countries. It was the governments of five oil-producing countries (Venezuela, Saudi Arabia, Kuwait, Iraq and Iran) that established the OPEC in 1960, in an attempt to assert more control over the activities of the large international oil companies that produced oil in their countries. The post-war oil-producing economies were increasingly dependent on the income from oil, but the government had no direct control over this income stream. The fact that the international oil market was dominated by a few international firms that captured most of the economic rents and profits (the lack of spin-off in the domestic economies) provoked governments into asserting their rights to a larger part of the economic rents. In the absence of a sufficiently large private sector, they aspired to redistribute this income in their economies through public investments.

The oil market developments in the late 1960s and early 1970s placed the oil-producing countries in a position to finally take over control over oil production from the large international oil companies, and, at the same time, secure more of the income stream for their domestic economy. Once the governments had secured their control over these income streams, they awarded (mostly newly founded) state companies by offering them a leading role in the economy. Transacting through a framework of public ownership of production was not limited to the oil industry, but was also implemented in other sectors of the economy. In part, the justification for such a dominant role of the state in the economy was, particularly for the Gulf producers, the lack of absorption capacity compared to the sheer size of the oil income, and the small size of the private sector in general. Governments had to regulate the infusion of capital in the economy in order to mitigate large distortions in factor allocation. The lack of a matured private sector, however small, only complicated managing the large income streams. Nonetheless, it cannot be the only explanation for the fact that the economy is organized in this particular way. Some of the OPEC member states had relatively diversified and mixed market economies and could have opted for regulatory controls rather than state ownership.

The grounds for the state's dominant role in the economy must also be found in the dominant post-war economic thought, the approach to development strategies and the political and economic order in the world. The role of the state in the economy in the post-1945 period was relatively large compared to other periods in modern economic history, and continued to grow until the late 1970s. In 1973, the oil-producing countries were in a position to establish a new organizational structure of their economies based on their massive oil incomes. Given the dominant ideas about management of the economy and the particular political make-up of most oil-producing countries, it was not surprising that the state used the oil incomes to take firm control over the economy.

The large redistribution of income from oil consumers to oil producers did not leave the world economy untouched. The massive balance-of-payments problems of oil-consumer countries, and the change in the cost structure of many companies as a result of the oil price increase, could not be solved within the structure of the 'old' economic order, nor with the tested economic policies of the previous decades. Eventually, after another round of oil price increases in the late 1970s, this led to a re-thinking of the role of the state in the economy. The world economic order is not a static concept, but a highly dynamic process that forces continuous changes on the organization of firms and governments alike. In this dynamic international economic situation, the role of the state in the economy in industrialized countries and later also the oil-consuming developing countries was, in the 1980s, changing from direct intervention to more regulated domestic economies. The transition in the late 1980s and early 1990s from a centrally planned economy to a market economy was the most dramatic example of a changing economic order. In the period that most other economies were redefining the role of the state in the economy, the oil-producing countries were engaged in a reversed process.

The process of increasing direct intervention in the oil-producing countries was stunted by international economic developments in the late 1980s and 1990s. The oil price fall, the depreciation of the dollar, and in some countries mounting debt problems in the mid-1980s forced the oil-producing countries to rethink their position. Slowly the oil-producing countries were beginning to shift away from direct intervention, but found it hard to implement an altogether new oil regime. The thrust of the liberalization process in the majority of oil-producing countries has so far been indecisive, and some proposed changes, such as in Indonesia, encounter opposition from various sections of society. However, any changes in this direction will have a major impact on the organization and operation of the national oil companies and, as a result, also on the structure of the international oil industry.

In the next sections we explore the shifting role of the state in the economy, liberalization and regulatory reforms, and the (international) economic events that reduced the ability of governments to intervene efficiently in the domestic economy. It will also be shown how international institutions, such as the International Monetary Fund and the World Bank, expedited the liberalization process in developing countries.

3.2 THE STATE VERSUS MARKET DEBATE

The evolution of the role of the state from providing 'law and order' as in policing to 'law and order' in an economic sense has been a long one. In addition to taxing, expenditure activities and state participation in production, modern states establish and enforce the rules under which the private economy can operate. They provide the legal enforcement of contracts, provisions for bankruptcy, laws defining property rights and liabilities. The state has a wide array of instruments at its disposal with which it can directly or indirectly influence the economy, such as direct controls, regulation, legislation controlling firms or unions, and monetary and debt policy.[39]

The debate on the degree of public involvement in the economy, first in relation to nationalization and more recently on privatization, and the related issues of the efficiency of markets, firms and government touches the core of economic theory controversies. The period 1945-1980 is strongly associated with the welfare state and the conviction that governments could address problems in the social and economic system, like in the Great Depression in the 1930s, to cure market failures. Market failures that are so serious that they require regulation, can be divided into three categories: problems of asymmetric information, problems of externalities and problems of monopoly power.[40] This intervention model is described as the

[39] A.B. Atkinson and J.E. Stiglitz, *Lectures on Public Economics*, McGraw-Hill International Editions, Economics Series, McGraw-Hill Book Company Ltd, London, 1980, p. 5.

[40] M. Armstrong, S. Cowan and J. Vickers, *Regulatory Reform, Economic Analysis and British Experience*, The MIT Press, Cambridge, Mass., 1994, pp. 11-3.

'helping hand of government' or 'public interest' model, in which the solutions to market failures ranged from corrective taxes, regulations, aggregate demand management, price controls, subsidies, planning and government ownership.[41] The 'public interest' theories emphasized the governments' role as a benevolent maximizer of social welfare both in the provision of traditional public goods and as an economic regulator. This approach has been criticized widely, because government interventions predominantly served the governments' political interests and rarely corresponded with social welfare, while the assumption that governments would maximize social welfare misguided policy-making.[42] Political economy theories of government challenged this view, asserting that interest groups in society have a stake in the outcome of the regulatory process, and actively seek to influence the formation of public policy.[43] These arguments further support the view that regulation can benefit industries and is often encouraged or controlled by these industries because they can gain from direct monetary subsidies, constraints or subsidies on substitutes or complements of the commodities produced by the industry, price fixing, and control over entry by new rivals.[44]

This last point, control over entry, is particularly interesting with regard to the oil industry. Governments auction exploration and production licences, and as often is the case, "the power to license becomes the license to exclude".[45] Moreover, in many oil-producing countries, the state oil company, operating as the government's agent, determines or biases entry, particularly through the administration of Production Sharing Contracts (PSCs) or through serving as a contractor with private international oil companies. As operators of PSCs, the NOCs can determine the quality of

[41] A. Shleifer and R.W. Vishny, *The Grabbing Hand, Government Pathologies and their Cures*, Harvard University Press, Cambridge, Mass., 1998, p. 2.

[42] J-J. Laffont and J. Tirole, *A Theory of Incentives in Procurement and Regulation*, The MIT Press, Cambridge, Mass., 1993, p. 475.

[43] P.C. Ordeshook, 'The Emerging discipline of political economy'. In J.E. Alt and K.A. Shepsle (eds.), *Perspectives in Positive Political Economy*, Cambridge University Press, Cambridge, 1990, pp. 14-20.

[44] J-J. Laffont and J. Tirole, *op. cit.*, 1993, p. 537.

[45] *Ibid.*

the business climate.[46] In Indonesia, Pertamina fulfilled this role. Interestingly, in the course of time the implementation of PSCs by Pertamina resulted in a negative business climate, and private international oil companies became increasingly reluctant to invest. In an attempt to improve the investment climate in the oil and gas sector in Indonesia, the minister for Mining and Energy in the Habibie government (1998-1999), Kunturo Mangkusobroto, proposed new legislation which would undercut the monopoly position of Pertamina, and proposed to open the sector also for private business entities. To date, this draft legislation has been stonewalled in parliament, due in no small part to active lobbying by Pertamina.[47]

The role of bureaucracies, in particular with respect to the agency relationship between politicians and their delegates in the bureaucracy, is important because political principals are at the mercy of better-informed agencies. Collusion with agencies or political principals in regulatory matters, depending on the level of independence or constraint of the agency, offer rent-seeking opportunities, which influence the outcome of the regulatory process.[48] This rent-seeking not only takes place within the industry itself, but also within the bureaucracy. This may not only lead to inefficient allocation outcomes, moving resources into established production or the public rent seeking-sector, but it could also lead to the imposition of taxes on innovative activities.[49] This could seriously hamper economic growth.[50]

The inefficiencies arising from corruption can be great indeed. In recent years, corruption ratings have been published to indicate the extent of the problems in many economies around the world. In the literature, a distinction is made between the effects of centralized and decentralized corruption. In general, centralized corruption is judged to be less harmful to efficiency than decentralized corruption. In the case of centralized corrup-

[46] T.N. Machmud, *op. cit.*, chapter 1 (to appear 2000).

[47] *Financial Times*, 13 July 1999. In Chapter 7, we will return to efficiency and corruption issues in NOCs.

[48] J-J. Laffont and J. Tirole, *op. cit.*, 1993, p. 501.

[49] K.M. Murphy, A. Shleifer and R.W. Vishny, 'Why is Rent Seeking so Costly to Growth?' In A. Shleifer and R.W. Vishny, *The Grabbing Hand, Government Pathologies and their Cures*, Harvard University Press, Cambridge, Mass., 1998, p. 89.

[50] A. Shleifer and R.W. Vishny, *op. cit.*, 1998, pp. 70-1.

tion, bribes can be combined in some sort of package deal, which does not distort too many decisions at the margin.[51] In a decentralized situation of corruption, various ministries, agents and/or local governments accept bribes in an effort to maximize their own revenue.[52] The latter can create a fragmented, secretive and fairly unpredictable system of corruption, while centralized corruption at least has the advantage of being predictable and orderly.

In this context, according to Bardham, Indonesia was an excellent example of more centralized corruption that was: "Controlled largely by the first family and the top military leadership in cahoots with the ethnic Chinese-run conglomerates".[53] The ousting of the Suharto regime in May 1998 may have ended centralized corruption or at least will have seriously impeded it. A transition to decentralized corruption could be disastrous for future efficiency and economic growth in a situation in which Indonesia is sliding towards a system with a weak central government. In this context, the outcome of the June 1999 elections is worrisome. Furthermore, with a reputation of corruption, it is hard to eradicate such practices for a lasting period of time.[54] That these worries are real is exemplified by the request of the IMF to perform an independent audit of three state companies in Indonesia (Perusahaan Listrik Negara, the state power utility; Bulog, the food logistics agency; and Pertamina the state oil and gas company). In an audit report by PricewaterhouseCoopers, which was leaked to the press, Pertamina was said to have lost $6.1 billion due to corruption and inefficiency in two years time.[55] The auditors found mark-ups of 30 to 36 per cent on charters, insurance premiums and chemicals; $400 million was lost on sales of natural gas below market price; $129 million in questionable fees to two oil trading companies (which are both partially owned by sons of

[51] P. Bardhan, 'Corruption and Development: A Review of Issues', *Journal of Economic Literature*, vol. XXXV (September 1997), p. 1325.

[52] A. Shleifer and R.W. Vishny, *op. cit.*, 1998, p.103.

[53] P. Bardhan, *op. cit.*, *Journal of Economic Literature*, vol. XXXV (September 1997), p. 1325.

[54] J. Tirole, 'A Theory of Collective Reputations (with applications to the persistence of corruption)', *Journal of Economic Studies*, vol. 63 (1996), p. 5.

[55] *Financial Times*, 13 July 1999.

former president Suharto) for services that Pertamina staff could easily have performed themselves.[56]

The evidence against the 'public interest' or 'helping hand of government' models gained momentum in the 1980s, when government policies clearly had not been able to produce the desired outcomes in a period of economic downturn. As argued earlier, the efficiency of factor allocation cannot be guaranteed by public intervention because government failures can also be large. Government failures include corruption, bribery, deliberate mis-allocation of resources, and also the organizational behaviour of bureaucracies, the lack of market discipline and accountability, and problems with information management.[57]

Although insights have been shifting, and early attempts to a less interventionist government appeared, the governments' role remained large and very important. However, evidence of the shifting role of the government in the economy was not reflected in the share of government spending in the GNP. In many countries, state involvement was large, and had been increasing over the years, particularly in terms of spending on social services (health and education), but also in terms of production. Despite the efforts of the OECD countries since the 1980s to reduce government expenditure, spending had increased to 50% of GDP in 1995, compared to about 45% in 1980.[58] In 1960, state spending in the OECD countries was about 20% of GDP. State spending in developing countries varied between 25 and 30% in the period 1975-1990. In 1960, spending was still 15% of GDP.

Governments not only increased spending on social services, they also became increasingly involved in production, either through subsidies or through state companies. State companies not only produced public goods, but were also active in utilities, primary energy (mining), steel, shipbuilding, banking, cars, aerospace, etc. In the 1980s, the role of the state and the involvement in production came under new scrutiny. Governments were burdened with large budget deficits, and, in some cases, failed to deliver

[56] *Financial Times*, 13 July 1999.

[57] D. Helm, 'The Economic Border of the State'. In *The Economic Border of the State*, D. Helm (ed.), Clarendon Paperbacks, Oxford University Press, 1990, p. 28.

[58] *The State in a Changing World*, World Development Report 1997, World Bank/ OUP, 1997, p. 2.

some basic public goods such as property rights, roads, health and education. Moreover, the governments' macro-economic policies no longer generated an internally balanced economy. As much as 'market failures' or 'private sector failures' had served as an explanation or justification for government intervention, the emphasis on 'government failures' had to justify the state's withdrawal from production. To be sure, 'market failures' do exist, as do 'government failures', but the market and the state are not each other's natural counterparts. The inefficiency of the market outcome does not necessarily lead to improvement when the government intervenes and *vice versa*.

The performance of the state as an entrepreneur in state companies is disappointing, particularly if they are held against certain welfare and production objectives.[59] The state companies were unable to adapt the production process to basic and developing human needs. More often than not, state companies are followers, and have little incentive to adapt to changing market conditions and when they attempt pioneering they make enormous mistakes. In times of crisis, state companies have turned out to be very inflexible in their capacity to restructure. Although state companies offer good employment conditions, it is hard to say if they also managed to achieve their social goals and contribute to a just distribution of welfare.

Given the fact that markets, firms and governments clearly fail to bring about optimal resource allocation, an important question comes to the fore: are these failures compensatory or are they supplementary. If they are compensatory, then the combination of 'market failure' and 'government failures' leads to the insight that only an optimal 'mix' of private and public sector involvement could lead to an efficient economic situation. Achieving the optimal 'mix' must take place in its dynamic context, only then a dynamic economic equilibrium is possible. Thus, following Atkinson and Stiglitz, "even if we accept the basic theorem of the efficiency of the competitive economy as a valuable reference point, there remain important reasons for government intervention. These may be summarized under the following headings: (1) distribution, (2) failure of perfect competition, (3) absence of futures and insurance markets, (4) failure to attain full equi-

[59] H.W. de Jong, 'The State as Entrepreneur: The rise and fall of an idea in the economic experiences of European countries'. In G. Aschinger et al., *Deregulierung -eine Herausforderung an die Wirtschafts- und Sozialpolitik in der Marktwirtschaft*, Duncker & Humblot, Berlin, 1989, pp. 186-8.

librium, (5) externalities, (6) public goods, and (7) merit wants."[60] If, however, the 'failures' are considered to be supplementary, no optimal resource allocation is possible, regardless of the public/private mix. Instead, there will be various equilibria and not a superior one. Since the state is more than a public producer, but is also a market and firm regulator, the state influences the efficiency outcomes for the firms and the markets. The regulatory powers of the state can cause both (intended and/or unintended) additional transaction costs and benefits, and there will be different 'trade-offs' between firms and markets.

The state does not have a superior position over firms or markets as far as obtaining symmetric information is concerned, and, as a result, regulation will also be imperfect. Vickers and Yarrow, modelling for asymmetries of information to find the optimal regulatory mechanism in these circumstances conclude: "In doing so, it illuminates the trade-offs between internal and allocative efficiency that result from asymmetric information, and it reveals how the effectiveness of regulation depends critically upon information available to the regulators."[61]

Vickers and Yarrow also consider the possibility of collusion between regulator and firm or government and firm when the government is 'captured' by firms "when designing regimes of competition and regulation for them".[62] Governments, also when running nationalized industries, have been known to intervene, but possibly also regulate, with their short-term rather than a longer-term (public) interest at heart. In relation to managing the nationalized industries, Vickers and Yarrow further state that: "(...) a system of control that relies heavily upon agents' internalization of public interest objectives is unlikely to produce good performance. In the event, as we have described above, the results of the policy failure have included widespread goal displacement, lack of clarity in corporate objectives, overlapping responsibilities, and excessive ministerial intervention in operational decisions. These, in turn, have had detrimental effects on pricing,

[60] A.B. Atkinson and J.E. Stiglitz, *op. cit.*, p. 8.

[61] J. Vickers and G. Yarrow, *Privatization, An Economic Analysis*, The MIT Press, Cambridge, Mass., 1988, p. 100.

[62] *Ibid.*, p. 109.

investment, and internal efficiency performance of the nationalized industries."[63]

The governments of oil-producing countries definitely founded the NOCs with mixed objectives. They very often meddle with the management of the company and also impose non-commercial demands.[64] Shleifer and Vishny go even further than that and claim that the inefficiency of the public enterprise is the "result [...] of political pressures from the politicians who control them."[65] Public ownership also introduces problems of monitoring and incentives. Multiple principals have comparatively weak incentives to monitor the activities of the agent, the managers of the NOCs, to make a credible commitment in pursuing the objectives of the principals. The structure and process of government relations in oil-producing countries allows managers of NOCs to pursue a range of objectives.

Efforts to commercialize the NOCs in the late 1980s and early 1990s failed because the governments reverted very quickly to their old behaviour. Empirical evidence further suggests that public enterprises (and NOCs are no exception) are over-staffed and pay more than market wages.[66] They are also located in politically desirable rather than economically desirable regions, and charge prices either significantly below marginal costs to win political support or monopoly prices when political objectives dictate this. Moreover, public enterprises do not deal with environmental externalities (market failure) very well because they are often the worst polluters (government failure). The evidence against public enterprises as a model to overcome market failures is large. The efficiency of public enterprises is lower than private enterprises, and in the case of the oil-producing countries, the economic rationality to organize their oil sector through an NOC is dwindling.

For the oil-producing countries, the desire to bring the assets of the international oil companies under state control was foremost fed by the long history of political and economic exploitation. In addition, the lack of

[63] *Ibid.*, p. 151.

[64] M. J. Horn, *The Political Economy of Public Administration*, Cambridge University Press, 1995, p. 139.

[65] A. Shleifer and R.W. Vishny, *Politicians and Firms*, Discussion Paper Number 1686, Harvard Institute of Economic Research, June 1994, p. 1.

[66] *Ibid.*, pp. 2-3.

domestic private initiative and/or the inappropriateness of private initiatives in the domestic economy gave the government another impetus to bring the oil industry under state control.[67] However, when the economic nationalistic aspirations were translated into the strong interventionist policies of the government, the new structure of the economy was much more an answer to the political and economic past rather than an anticipation of the economic future of the country. Although the government had to assert its political and economic rights under the circumstances, the resulting structure of the economy did not reflect much thought on problems of economic efficiency and the drawbacks of direct intervention on the efficiency of the firms. It is therefore not surprising that within 20 years of the nationalization of the oil industry, the boundaries between state and firm need to be redefined.

3.2 The State versus Market in Developing Countries

The debate on the welfare state and the role of the state in the economy has also dominated the debate on development and strategies of developing countries. The convergence predicted by mainstream economic theories on the competitive market, where trade, capital, and migration flows are supposed to equalize returns to factors of production and diminish national differences, proved to be inconsistent with practical experience.[68] In reality, national differences increased rather than diminished, despite the intensification of international economic interaction in the last century. For developing country policy strategists this posed questions with regard to the desired level of integration in the international economy and the role of the state in their domestic economies.

Nevertheless, the post-World War II economic order was founded on these mainstream insights. The Bretton Woods institutions and the GATT were established to achieve free trade and to offer funds to governments to finance temporary balance-of-payments deficits and structural investment. Although a discussion about the suitability of these institutions to achieve equalization of returns to factors and economic convergence between natio-

[67] J.-P. Anastassopoulos, G. Blanc and P. Dussauge, *State-owned Multinationals*, Wiley/IRM series on Multinationals, Wiley & Sons, Chichester, 1987, p. 33.

[68] A. Fishlow, 'Inequality between Nations.' In *Economic Development*, J. Eatwell, M. Milgate, P. Newman (eds.), The New Palgrave, The Macmillan Press Ltd., London, 1987, p. 176.

nal economies would take us too far[69], the perception of developing countries that the international economic order was asymmetrical and not beneficial to their development greatly influenced their national strategies. Theories like the Singer and Prebisch thesis of a secular decline in terms of trade of primary products held that through declining terms of trade, unequal exchange, technological dependence and domestic disintegration, national economies would diverge rather than converge. The structuralist school argued that socio-political structures in the world "affected, even shaped the market, and thereby in (large) part determined advantage in the market. The industrialized countries had 'self-sustained' economies whereas the underdeveloped countries did not, since they functioned as peripheries to centers. The world market forces operated to maintain this undesirable 'equilibrium'. The policy implications were clear. Since the 'normal' operations of the market would only continue the same pattern, state action was required to alter it."[70]

The 'structuralist' theories strengthened the belief in inward and interventionist industrialization strategies of most developing countries' governments. Thus, the structuralist approach to trade varied widely from the neoclassical approach, but they agreed on the importance of trade to economic development. According to Tussie, the structuralists were "merely advocating a recasting of principles of international trade in a Keynesian light. Unregulated market forces could not be relied upon to solve inequalities between countries, just as they could not do it within countries."[71]

Inspired by the post-1945 reconstruction experience in Europe, the new concept of economic development of in developing countries became synonymous with industrialization. However, the support of government to the industrialization process cannot alone explain the diverging progress in industrialization. Political, social and structural factors are just as important. Kumar Bagchi asserts that "It is doubtful whether a general pattern of successful state intervention to overcome economic backwardness can be discerned from historical experience as has sometimes been claimed. What

[69] But see for instance, D. Tussie, *The Less Developed Countries and the World Trading System, A Challenge to the Gatt*, Frances Pinter, London, 1987.

[70] I. Wallerstein, 'Periphery'. In *Economic Development*, J. Eatwell, M. Milgate, P. Newman (eds.), The New Palgrave, The Macmillan Press Ltd., London, 1987, p. 271.

[71] D. Tussie, *op. cit.*, p. 22.

can be asserted is that state intervention in industry was much more likely to succeed in countries where capitalist relations had advanced far than where intervention from the top was used as a substitute for social change which might upset the balance of class forces among the rulers."[72]

In developing countries, state intervention was further inspired by the lack of a modern private industrial sector that could spearhead the modernization of the economy and increase economic growth and welfare. The private sector in developing countries is usually considered too small, and could not generate sufficient capital requirements for economic growth in the early stages of development. The weakness of the internal banking system impeded investments. In addition, in developing countries, state involvement in economic activity is invariably explained by market failures, where "Social overhead capital, or infrastructure, typically enjoys substantial economies of scale, is lumpy, must be coordinated with other location-specific investments, and may be difficult or inefficient to supply through the market. (...) If planning is to be consistent, it will typically attempt to influence private decisions, and many of the distortions (...) may be traced to attempts to control or direct private sector decisions in directions favoured by the plan."[73]

A strong government was seen as instrumental in achieving the developmental goals. The economic and political ideology of the time in the OECD countries merged with the interests of the political leadership in developing countries. Many of the developing countries as we know them now had or were about to become politically independent states. This was achieved with much effort. However, the strong economic ties with their parent countries were not so easily severed. More often than not, the former parent country was still the source of 'advice', economic aid and planning. The fact that many of the leaders of the developing countries were trained in the educational systems of the former parent countries and, as a result, also trained in the predominant economic theories, facilitated an easy transfer of ideas on economic development and management to the elites in deve-

[72] A. Kumar Bagchi, 'Industrialization' In *Economic Development*, J. Eatwell, M. Milgate, P. Newman (eds.), The New Palgrave, The Macmillan Press Ltd., London, 1987, p. 164.

[73] D. Newbery, 'Taxation and Development'. In *The Theory of Taxation in Developing Countries*, D. Newbery and N. Stern (eds.), World Bank/Oxford University Press, 1987, p. 175.

loping countries. In addition, the economic institutions in the developing countries were erected and crafted by the former parent countries, and continued to influence policies after independence was achieved. Hettne observes that "It is thus somewhat paradoxical, but not inexplicable, that of those development strategies which have emerged from Western historical experiences, the one that in retrospect appears as least relevant -the Keynesian- came to be of most immediate importance for the birth of development theory. The reason was of course a coincidence in time (the late 1940s) between Keynesian theory as a predominant school in economics and the growing interest for 'backward areas', by which name the underdeveloped world was known in those days. The character of the modern elite and its close relations to the state made strategies which emphasize the role of state power more attractive than purely liberal alternatives. The Western model was thus not in any way forced upon the developing countries. It was fully consistent with the power structures in those countries."[74]

The changing role of the state in the industrialized countries, combined with the internationalization of the economy, had to have an impact on the role of the state in developing countries. It therefore comes as no surprise that from the 1980s onward, the role of the state in developing countries is also changing, sometimes under pressure from international organizations like the IMF.

3.3 Internationalization of Production and Capital

The biggest lacuna in the recent debate on State versus Market, however, has been the effect of internationalization of the economy on the economic powers of the state. Welfare economic goals must take place in a dynamic environment in which the economic policies and production decisions of the state, but also of the market and firms must function and adapt. The growing interdependence of national economies has invalidated certain tools to achieve macro-economic equilibrium, but the policies did not reflect these changes sufficiently. The changes in the international economic order as they were laid down in the Bretton Woods monetary rules and the GATT trading system and the expansion of international production and capital markets, which effectively undermined national economic policies, have radically changed the position of the state as an effective

[74] B. Hettne, *Development Theory and the Three Worlds*, Longman Development Studies, Longman Scientific & Technical, New York, 1990, p. 49.

economic macro-economic regulator.[75] The limitations of the old economic order became apparent in the early 1970s, and while the institutions slowly tried to adapt to world economic events, the international economy experienced a period of severe factor price instability. This instability affected trade, industrialization patterns, capital flows, and employment opportunities, which signalled new competitive advantages for some, and declining ones for other nations. However, there was also a certain perversity to the signalling, which led to overambitious investments in, for instance, the oil-exporting countries. The uncertain economic environment of the 1970s and 1980s resulted at first in more state intervention in the industrialized and developing countries, to be followed by a period in which governments, realizing their inability to moderate the changes taking place, revised their policies and tried to scale down intervention in the (late) 1980s. Instead, industrialized countries increasingly opted for privatization, deregulation and regulatory reforms. However, this policy U-turn took place without a new 'economic order' in place, and without a clear idea of how to address the instability in the world economy. The institutional imperfections of the international institutions (IMF, World Bank, and GATT/WTO) became more and more apparent in the 1980s and 1990s, when international economic instability increased. The internationalization of capital made all markets more vulnerable to instability because of economic interdependency. Furthermore, the growing recognition of environmental problems, and especially the transnational nature of global warming, has contributed further to the awareness of the limitations of national action programmes *versus* international measures.[76] Put differently, macro-economic regulation had developed a need to be internationalized parallel to the economy, but the institutional make-up of the international economy was incompetent for this task.

Public sector involvement in economic activity has been growing in all economies after 1945. In the industrialized countries, the welfare state gave

[75] L. Thurow, 'A New Economic Game'. In *International Political Economy, State-Market Relations in the Changing Global Order*, C. Roe Goddard, J.T. Passé-Smith, J.G. Conklin (eds.), Lynne Rienner Publishers, Boulder, 1996, pp. 130-45.

[76] H. French, 'Forging a New Global Partnership'. In *International Political Economy, State-Market Relations in the Changing Global Order*, C. Roe Goddard, J.T. Passé-Smith, J.G. Conklin (eds.), Lynne Rienner Publishers, Boulder, 1996, pp. 479-508.

rise to more state intervention on both the micro- and the macro-economic level.[77] However, the conception of the welfare state was based on a mixture of economic theories and moral and political ideas and practices. The Keynesian insight that economies were not necessarily self-correcting and that demand could be managed by government, and the socialist idea of economic planning, for instance, initiated a period of increasing government intervention in all national economies. Moreover, the macro-economic stabilization policies became a necessary part of implementing the international institutional arrangements on trade (GATT) and exchange rates (IMF/Bretton Woods), which had been established to guarantee stable international trade and monetary relations in the world, thus linking the national intervention policies to the international economy.[78] The governments' fiscal and monetary policies were tools to equilibrate internal and external balances. In a substantial number of industrialized (European) countries, governments also brought 'basic industries' under state control. The energy industries were significant among them. At the time, the argument for state ownership of these industries was that nationalized industries were part of the instruments for economic management, that planning was considered superior to market allocation of resources, and that nationalization served industry-specific purposes to do with monopolies, social costs and cross-subsidization.[79] The policy aim of full employment and workers' desire to have more control over production converged in state control of industries. Helm comments on the nationalization of industries: "But the moral and political commitment lacked an instrumental basis -the practical aspects of implementing this vision were less clearly spelt out. In reality, many of the important caveats of the economic theory were ignored and a series of ad hoc measures employed. Thus market failure arising from monopoly and collusion was 'solved' by nationalization, without due attention to 'government failure' and hence little careful attention to the form of controls."[80] In

[77] A. Santos, 'Privatization and State Intervention - (An Economic Approach)'. In *Deregulation or Re-regulation?, Regulatory reform in Europe and the United States*, Giandomenico Majone (ed.), Pinter Publishers, London, 1990, p. 139.

[78] In the debate State versus Market, there was little attention to the international dimension of economies and the change in the post-war international economic order.

[79] D. Helm, *op. cit*, p. 29.

[80] *Ibid.*, p. 27.

the 1970s, when the economic downturn in the labour-intensive sectors of the economy of the industrialized countries threatened employment, yet more industries were brought under state control.

In the 1980s, governments, in particular in the OECD but also progressively in developing countries, started to withdraw from the economy as producers. Governments recognized that the 'market' was a more efficient tool for resource allocation, and their role should be limited to regulation of the market. This change in philosophy was inspired both by theoretical and practical concerns.[81] The supply-side and monetary-economic theorists provided the theoretical foundations for this change, admittedly rather weak in the case of the supply-side approach, while budget restraints, structural balance-of-payments problems and the phenomenon of stagflation provided the practical grounds for governments to redefine their role in the economy. The renewed debate on State versus Market was thus as much a result of failing government policies, as of new insights into the economy. In De Jong's words, "Hardly anyone promotes the role of the state for ideological reasons anymore, and those few who do are on the defensive."[82] However, the conclusion that the government must now give way to the market because the market can correct the 'government failures' commits the same "functionalist fallacy of assuming that the logical existence of a role for the state can explain why it came into being and behaves as it does."[83]

The internationalization of macro-economic stabilization policies is the result of growing economic interdependence of national economies. The mobility of capital and the existence of international capital markets, which are not within the reach of national regulation, have curbed the possibility for governments to implement independent stabilization policies such as interest rate changes and other demand-switching policies. International co-ordination of policies is considered necessary, but the loose structure of the

[81] P. Krugman, *Peddling Prosperity, Economic Sense and Nonsense in the Age of Diminished Expectations*, W.W. Norton& Company, New York, 1994, pp. 55-103.

[82] H.W. De Jong, *op. cit.*, p. 181.

[83] A.B. Atkinson and J.E. Stiglitz, *op. cit.*, p. 9.

coordination, as in the G-7[84], has so far not been able to prevent policy competition among states.

3.4 REGULATION, DEREGULATION, RE-REGULATION

When in the late 1970s and early 1980s many governments in the Western world came to the conclusion that the management of the productive sector was ready for a re-think, the pressure on the performance of both public and private firms had become very great. Large changes in factor prices, and monetary and macro-economic instability made old policy recipes inadequate to resume sustained growth paths. International competition was increasing after the Tokyo Round tariff reductions, and the internationalization of production and capital markets were increasing in pace. The realization that the state as a producer of certain goods and services underperformed the competitive market initiated a process of liberalization and privatization. Simultaneously, the ways in which governments regulated the private sectors also came under scrutiny, and started a process of regulatory reforms. This process did not take place simultaneously in all Western economies, and is still ongoing.

At the same time, governments in fast growing developing countries reaffirmed their control over the economy, either through direct involvement or public firms, strict regulation of the private sector or through what since 1997 has been known as, "crony capitalism".[85] Wade[86] concludes that the government played an important role in the successful growth in East Asian in the 1980s and early 1990s: "But the difference between what happened in East Asia and elsewhere does not lie in the discovery of industrial policy instruments not known elsewhere. Many other nations have at one time or another tried most of the policy tools in East Asia. What

[84] Group of seven major industrialized countries: United States, Canada, Japan, Germany, United Kingdom, France and Italy. Lately, Russia has participated in the discussions of the Group-7. On certain issues only 5 countries meet in the Group of 5.

[85] Crony capitalism became a popular term during the Asian Financial Crisis of 1997-1998 in which the private sector was dominated by persons closely associated to the government and in which these private firms were often awarded monopolies. Elite capitalism would be a good synonym.

[86] R. Wade, *Governing the Market, Economic Theory and the Role of Government in East Asian Industrialization*, Princeton University Press, Princeton, 1990.

differentiates their efforts, above all, are a consistent and coordinated attentiveness to the problems and opportunities of particular industries, in the context of a long-term perspective on the economy's evolution, and a state which is hard enough not only to produce sizable effects on the economy but also direction of the effects, which is a more demanding achievement."[87] The much heralded economic growth strategy of East Asia was therefore not just the success of the market, as neo-classical theorists claimed, nor entirely government-led, as political economists claim, but due to governing the market. Wade continues in his concluding chapter with ten policy prescriptions of which the gradual liberalization of trade and the recommendations of a bank-based financial system (like Germany and Japan) as opposed to the Anglo-Saxon model of market-banks stand out.[88] He argues that a cleavage between the domestic financial system and the international financial system, and thus some government control, is vital for the development of the economy.[89] Unfortunately, the Asian crisis has redirected regulatory reforms. Although, the IMF underwrites the notion that liberalization needs to take place gradually[90], and that domestic liberalization comes before external liberalization, they are exerting great pressure on countries in their stabilization programmes (notably South Korea, Thailand and Indonesia) to quickly implement the necessary internal regulatory reforms (competition policy, bankruptcy laws, supervision of banks, etc.) in order to integrate into the world economy. Thus, the type of liberalization and regulatory reform presently underway is of the 'IMF type' rather than the 'Wade type'.

The State versus the Market debate, both in industrialized and in emerging market economies, is very much a debate about regulation and/or regulatory reform, of which the latter is also discussed in conjunction with the privatization of public firms and the introduction of competition. In

[87] *Ibid.*, p. 343.

[88] For an assessment of the respective corporate governance systems see J. Macey and G. Miller, 'Corporate Governance and Commercial Banking: A Comparative Examination of Germany, Japan, and the United States', *Stanford Law Review* 48, 1995, pp. 73-112.

[89] R. Wade, op. cit., pp. 366-8.

[90] M. Camdessus, *Capital Account Liberalization and the Role of the Fund*, IMF, Washington D.C., March 9, 1998, p. 2.
(Http://www.imf.org/external/np/speeches/1998/030998.htm)

most oil-producing countries, a national oil company (NOC) was established to carry out all activities related to the oil industry in the country. The governments therefore opted to monopolize the sector or, in countries where private (international) firms were allowed a certain presence in the oil sector, the NOCs were given market power. In some countries, the NOC was the only operator in the sector, whereas in other countries, the NOC always had a majority stake in a project, and/or arranged PSCs. The activities of foreign investors in the oil sector were with a few exceptions, restricted to exploration, production and the export of crude. In some countries, such as Saudi Arabia, foreign oil companies were allowed to invest in the petrochemical sector in joint venture with a state company. Domestic distribution of oil products like petrol is reserved for the NOC. This monopoly in fuels is a blessing in disguise. In most oil-producing countries, the NOCs were forced by the government to supply fuels below cost, which constituted a major hemorrhage on the company's profits. Because monopolization of the sector was not enough, the governments also chose to strictly regulate the NOCs on the domestic market. It is therefore not surprising that the performance of these NOCs was very disappointing indeed, and that after 1986, liberalization and/or privatization of the oil sector was considered more seriously. Although countries have been very slow in implementing these regulatory reforms, the pressure to do so is mounting. The continued weakness of the price of oil in the 1990s has put the governments' budgets under great pressure and has further frustrated the performance of the NOCs.

Of course, liberalization and/or privatization alone cannot address the problems of underperforming NOCs. To this end, competition and regulatory reforms must be introduced. The sequencing of privatization, liberalization and regulatory reforms poses some difficult dilemmas.[91] Delay of privatization imposes substantial economic costs. First, money-losing state companies may continue to drain money from public finances, and second, the anticipation of privatization can entice the state company's management and/or workers to steal the most valuable assets. Last, the underperforming state company can obstruct liberalization and restructuring in other sectors of the economy. If privatization is preceded by liberalization and regulatory reforms the economic benefits will be greater. However, delay of privatiza-

[91] *The State in a Changing World*, World Development Report 1997, World Bank, Washington D.C., p. 63.

tion can cause management of a state company to become badly entrenched.[92] Despite the sequencing benefits, the World Bank admits that the political situation in many countries demands that privatization is kept on the front-burner, and that the disadvantages of early privatization, before liberalization and regulatory reforms have been fully introduced, should be preferred. In the oil industry this would create a situation in which newly privatized firms would be in a position to divert too large a portion of the economic rents away from the government. The break-up of the state oil company's monopoly in oil reserves and production assets, the introduction of a concurrent tax regime and, the introduction of a competitive bidding system for exploration areas could prevent this undesirable diversion of economic rents from the state. This would plead for building up at least some institutions that begin to regulate the market and that protect ownership rights. Once private sector activities are allowed to increase, private owners will have a incentive to lobby the politicians to step up the regulatory reforms.[93]

The liberalization and privatization of the Russian oil industry is a perfect example of incomplete regulatory reforms.[94] The resulting inefficiencies and mis-allocation of resources are so great that one would wish to go back in time and try again. The break-up of the state oil assets in several regional oil companies created regional monopolies. Prices at which the companies were sold were too low because privatization was restricted to predominantly Russian capital. Foreign investors were allowed only a modest interest in the initial rounds of privatization. Due to a lack of properly developed internal capital markets, private investment capital was scarce and in the hands of insiders, who managed to carve out good deals for themselves. Furthermore, competition on the internal market was non-existent, and prices were set (low) by the government rather than by the market. Formally, international competition was not allowed until November 1997, but even now domestic prices are still controlled by the government. The Russian oil industry remained strongly biased against production for the domestic market, and the government had to regulate exports strictly to

[92] *Ibid.*

[93] A. Shleifer and R.W. Vishny, *op. cit.*, 1998, pp. 149-50.

[94] M. Messer, *The Reorganization of the Russian Oil Industry*, Draft Doctoral Thesis, Leiden University, to appear in 2000.

overcome this bias. Since 1990, Russian oil production has declined dramatically, and yet it has managed to step up exports. This has partly been due to a dramatic decline in oil consumption and a near total collapse of the refining industry. Russia joined the ranks of the oil-producing countries that have to import products like gasoline. Russia's big mistake was not to introduce competition on the domestic market, to separate the liberalization and privatization of oil production from oil processing, and not to allow foreign investors to input capital and technology in both exploration, production and processing of oil. The Russian government sold off the 'crown jewels' of the oil industry first, and is now left with a processing and transportation sector (mainly pipelines) in total disarray. Meantime, the government has failed to put a proper regulatory regime in place that creates enough investment incentives in the oil industry and that will, at the same time, generate sustainable tax income from the oil.

The form of competition introduced depends strongly on the type of regulatory intervention required to correct market failures. The efficiency of relatively weak forms of competition (yardstick competition, public sector competition, capital market competition and competitive franchising), often considered as a transitional stage on the way to competition from exiting firms, new firms created by privatization, entry and international competition, is not convincing.[95] The introduction of stronger forms of competition may also require regulatory intervention. Problems of interconnection pose a serious problem for oil-producing governments that are considering introducing competition. Pipelines, for instance, are an important asset of the company that owns them, and market entry could be impeded if access to these pipelines is limited by the privatized NOC. The pipelines are natural monopoly segments of the industry to which rivals in complementary segments must have access in order to be able to compete.[96] Setting access charges would be a solution to the problem of introducing competition.[97] The oil industry is a particularly vertically integrated industry, both on domestic markets and international markets. A radical solution

[95] M. Cave, *Recent Developments in the Regulation of Former Nationalised Industries*, Government Economic Service Working Paper, No. 114/ Treasury Working Paper No. 59, London, 1991, pp. 9-13.

[96] M. Armstrong, S. Cowan, J. Vickers, *op. cit.*, pp. 135-49.

[97] J-J Laffont and J. Tirole, Creating Competition through Interconnection: Theory and Practice, Department of Economics, MIT, 1994, Mimeo, p.1.

to the problem of avoiding natural monopoly segments would be to force the dominant firm (the former NOC) to break up. However, this would ignore the economies of scale and scope. The organization of the oil industry tends to vertically integrate, and international competition would be impossible for the former NOC if efficiencies of vertical integration were denied. However, dominant market positions should be prevented by the implementation of proper anti-trust legislation. It would be much better to create a competitive market in which the former NOC must bid for investment projects (ranging from exploration to petrol stations) with other (possibly foreign) companies, and to allow the former NOC also to invest abroad. For the natural monopoly segments of the industry, access charges should be set up, and a regulatory board must oversee the pricing system.

Private shareholders will reward the former NOC for efficiently allocating capital by investing more in the company. Governments, on the other hand, tend not to reward good firm performance, because they pursue different objectives. Furthermore, the government might actually prevent good performance of NOCs because they starve the companies of capital.

3.5 TURNING POINT

The oil price increase of 1973 was very much a component of the increasing economic instability in the world. In March 1973, the Bretton Woods Exchange rate system collapsed, and in October oil prices started to rise. The inextricable political and economic events of that year continued to reverberate through the years that followed, when governments and international economic institutions tried to grapple with the large balance-of-payments instability in the world. The large shift in relative prices as a result of the monetary crisis and the oil price crisis created a significant realignment in the competitive advantages of countries. The ensuing reallocation of production factors in the world also involved the oil-producing countries, which appeared to have gained advantages in capital and energy-intense industries. However, maintaining the perceived competitive advantages was extremely difficult in the ongoing situation of international economic instability. The inability (or disinclination) of the governments of the leading industrial nations to reinstate the old economic order, and the increasing (protectionist) competition among the large trading blocks, changed the world order.[98] An order in which the oil-producing countries were trying to carve out a place for themselves as industrialized countries.

Before considering the achievement of governments of the oil-producing countries in absorbing the large windfall profits and before considering the contribution of OPEC to stabilize the international oil market, it is important to recognize the impact of the post-1973 international economic instability on the reach of development strategies. Particularly the fluctuations of the dollar and, therefore, international and US monetary policy are important for the oil-producing countries.

[98] R.E. Baldwin, 'Adapting the GATT to a more Regionalized World: A Political Economy Perspective.'In K. Anderson and R. Blackhurst, *Regional Integration and the Global Trading System*, Harvester/Wheatsheaf, New York, 1993, pp. 387-407.

CHAPTER 4

International Economic Instability

4. 1 THE CHANGING INTERNATIONAL ECONOMIC ORDER

Until 1973, the coordination of policies was institutionalized in the Bretton Woods exchange rate system. The fixed exchange rate system tied all currencies to the dollar, and the dollar to gold. Maintaining the exchange rate implied a balance-of-payments equilibrium, and stable or at least non-diverging developments in factor price. Free trade and the ability to equalize factor prices across national borders was one of the cornerstones of the system and was supported by the exchange rate regime of Bretton Woods. However, in a world made up of diverging economies in terms of structure and level of development, the stability of the exchange rate rested on rigorous stabilization policies of the member states and the flexibility of the system to incorporate changing production and trade patterns.

The Bretton Woods system was not faultless, nor, with the benefit of hindsight, timeless in its set-up.[99] The IMF was empowered to dictate stabilization policies to national governments when these governments had to draw on the higher tranches of the Fund for balance-of-payments assistance. In practice, the IMF only had powers to impose conditionality on economies with deficits, and only when these deficits were structural.[100] In fact, the IMF could not exercise such authority when a country's balance of payments was structurally in surplus, which was in principle just as destabilizing as a deficit, because such a member state did not need to draw on the higher credit tranches. Moreover, the discipline of the system relied on the continued strength of the dollar, and thus on a structural surplus in the American balance of payments. However, the system also relied on dollars being available abroad. The dollar was both intervention and reserve

[99] P. Krugman and M. Obstfeld, *International Economics, Theory and Policy*, Addison-Wesley, Reading, Mass., 4th edition, 1997, pp. 548-67; M. Obstfeld and K, Rogoff, *Foundations of International Macroeconomics*, MIT Press, Cambridge, Mass., 1997, pp. 567-69. See also S. Strange, *States and Markets, An Introduction to International Political Economy*, Pinter Publishers, London, 1988, pp.102-9.

[100] J. Williamson and C. Milner, *The World Economy, A Textbook in International Economics*, Harvester/Wheatsheaf, New York, 1991, pp. 378-80.

currency for the system and required the other countries to have enough dollar reserves to intervene in the currency markets. These countries could only obtain these dollars by creating a surplus in trade or capital flows with the United States or trade their goods, such as oil, in dollars. Because every currency was linked to the dollar, there was a lot of pressure on the United States to supply enough dollars on the international market. Worries about a liquidity problem (which was labelled the *Triffin Dilemma*[101]) led eventually to the creation of Special Drawing Rights (SDRs). Towards the end of the 1960s, confidence in the value of the dollar was evaporating and dollars were exchanged for gold.[102] The persistence of balance-of-payments problems among the industrialized countries, despite the expansion of the world economy and world trade, was an indication that adjustment problems had to be solved by adjustments in the exchange rate mechanism.[103] However, there was no agreement on reforming this system, and the American balance of payments deteriorated in the late 1960s.[104] The American deficit was offset in surpluses in the other industrialized countries, and also reflected growing exchange rate disparities. Despite the growing monetary uncertainty in the early 1970s, the industrialized countries showed little enthusiasm to share the responsibility for the adjustments problems in the Bretton Woods system with the United States.[105] The dichotomy between domestic monetary requirements and international requirements was mounting. The system had implicitly given the United States discretionary power to influence relative prices in the world, and, furthermore, it enabled the US government to export domestic imbalances. The inherent paradoxical conditions required for the system to function precipitated its downfall in the early 1970s.

[101] *Ibid*, p. 381. Named after Robert Triffin, *Gold and the Dollar Crisis*, Yale University Press, New Haven, 1960.

[102] C.P. Kindleberger, *A Financial History of Europe*, Oxford University Press, Oxford, 1993, p. 453.

[103] M. Garritsen de Vries, *Balance of Payments Adjustment, 1945 to 1986, The IMF Experience*, International Monetary Fund, Washington D.C.,1987, pp. 89-91.

[104] *Ibid.*

[105] I.S. Friedman, 'The International Monetary Fund: A Founder's Evaluation'. In *The Political Morality of The International Monetary Fund*, R.J. Myers (ed.), Ethics and Foreign Policy, volume 3, Transaction Books, New Brunswick N.J., 1987, p. 17.

The lack of IMF authority to discipline American economic policy (and most of the surplus countries) on which the system was so dependent brings us to another major hiatus in the system. The increasing over-valuation of the dollar in the late 1960s, for instance, inflated factor prices denominated in dollars. This led to distortions in the balance of payments among countries, and the misallocation of factors among economic sectors.

For a product as important to the national economies and their competitiveness as energy alone, such distortion is serious. Despite the price administration system of the large international oil companies in the 1960s, oil prices on the international market were relatively low as a result of the voluntary isolation of the American oil market from the international oil market and the of increased supply from new entrants in the market.[106] Downward pressure on the price and the declining purchasing power of the dollar increased the pressure to produce more oil. Oil is traded in dollars and the dual downward pressure on oil income reduced the oil-producing countries' ability to finance imports and investments. Although demand was expanding rapidly in the 1960s, the oil-producing countries themselves did not have any discretion with respect to the volume of production. Their production levels, which determined revenues, were administered by the international oil companies and were subject to power politics as much as to company production strategies.

The oil-producing countries' income depended largely on the volume produced in their countries and, given the ample supply on the market, the revenues that were due to them depended much on the political and economic importance of the country in question. This is particularly true for the consortia in the Middle East, which were balancing production in the region in such a way that it reflected the importance of these countries to American foreign policy.[107] Since the companies were also balancing the flow of oil on the world market in accordance with the growth of demand, it was at their discretion which oil to produce and in what volume. Although the participating companies in the consortia did not have complete parallel interests as far as production levels were concerned, the long-term contracts could satisfy the needs of those companies that were crude short. In this way, the political and economic demands made on the large international

[106] C. van der Linde, *op. cit*, 1991, pp. 125-42.

[107] *Ibid.*, pp. 120-21.

oil companies were served without upsetting supply and demand conditions. The competition from new entrants in the market made their balancing task harder and resulted in an erosion of the posted prices. The oil-producing countries, however, faced a market structure in which the distribution of oil revenues was not an expression of their oil reserves, nor of their production capacity potential, nor of the competitiveness of their oil.[108]

The misallocation of factors as a result of the international monetary system was much more widely spread than in the energy industries alone. The increasing pressure on the dollar in the late 1960s and the early 1970s finally led to the breakdown of the Bretton Woods system in March 1973[109], when confidence in the dollar and American economic policy had evaporated. The organization of the international oil market was slowly changing too. The dominant market position of the international oil companies was challenged, on the one hand by new entrants in the market and, towards the end of the 1960s, on the other hand by the oil-producing countries organized in the OPEC. The fall of the Bretton Woods exchange rate system, the rise of OPEC as an international oil market regulator, the growing importance of international trade and foreign direct investments, and the political and economic emancipation of the developing countries formed part of the changing international economic relations and the further internationalization of the markets.

4.2 Macro-economic Instability and Development

The Bretton Woods fixed exchange rate system was initially replaced with a system of floating exchange rates, in which some member states of the European Communities decided to float together. However, the oil price

[108] B. Shwadran, *The Middle East, Oil and the Great Powers*, Atlantic Press, 1955, p. 343.

[109] In August 1971, President Nixon suspended the convertibility of the dollar into gold and an across the board import tariff implemented to force, in particular, European countries to revalue against the dollar in exchange for a devaluation of the dollar. In December 1971, the Smithsonian Agreement gave the Bretton Woods system a new lease of life. The convertibility of the dollar into gold was not restored, and a *de facto* paper standard was established. However, the American balance of payments failed to recover under the new exchange rate regime and the system collapsed already in March 1973. M. Garritsen de Vries, *op. cit.*, pp. 90-4.

increases of 1973-1974[110] created such severe balance-of-payments difficulties for the European countries that a substantial number of participating currencies had to be withdrawn from the 'Snake-arrangement'[111] and had to adopt floating exchange rates too. Only in 1978, when the economies had recovered from the oil price increase and the ensuing recession, was exchange rate stability in Europe restored, and the European Monetary System was established.

The sudden increase in oil prices further distorted the balance-of-payments situation of countries.[112] The short-term inelasticity of the demand for oil was translated into large deficits in the trade balance. The quadrupling of oil prices prompted a transfer of income from oil consumers to oil producers that potentially could exert a large deflationary pressure on the world economy if this income was not spend.[113] The ability of the oil-producing countries to spend was, however, limited. Initial fears about large

[110] D. Yergin, *The Prize*, Simon & Schuster, New York, 1991, pp. 588-612. Oil prices increased as a result of an oil embargo against the United States, the Netherlands and later Portugal by the OAPEC (Organization of Arab Petroleum Exporting Countries) at the start of the October War in 1973, and the announcement to cut production as long as Israel occupied the Westbank and Gaza. The cuts were never implemented, but the perceived scarcity on the world market and the new sovereignty over oil reserves and production by OPEC member states encouraged OPEC to raise oil prices.

[111] B. Eichengreen, *Globalizing Capital, A History of the International Monetary System*, Leuven University Press, Leuven, pp. 152-57. In July 1972, the 'Snake arrangement' was established among the European Communities' member states and in which some other non-EC European countries participated, such as Norway, Sweden and Austria. The participating currencies could not fluctuate more than 2.25% in relation to each other. In the Washington Agreement of December 1971, the fluctuation margins of Bretton Woods were widened to 4.5% in relation to the dollar. In June 1972, the British pound left the 'snake' and floated from then on. In January 1974, France followed, re-entered the 'snake' in 1975 and left again in 1976. Denmark also left the 'snake' for a few months in 1972. In 1978, Germany, the Netherlands, Belgium, Luxembourg and Denmark participated in the 'snake'.

[112] B. Eichengreen and P.B. Kenen, 'Managing the World Economy under the Bretton Woods System: An Overview'. In P.B. Kenen, *Managing the World Economy*, Institute for International Economics, Washington D.C., 1994, p. 39.

[113] See for a discussion of the transfer problem R.E. Caves, J.F. Frankel, R.W. Jones, *World Trade and Payments, An Introduction*, HarperCollins College Publishers, New York, 6h edition, 1993, p. 63 and pp. 181-2; M. Obstfeld and K. Rogoff, *op. cit.*, pp. 255-6.

transfer problems proved unfounded, when a large part of the incomes re-entered the monetary system through the international private banks. This phenomenon is known as the recycling of petro-dollars, which contributed to the development and transformation of the international capital markets.

From 1974 until 1981, inflation escalated and balance of payment imbalances increased in most industrialized countries as a result of the inflationary fiscal policies of governments. The lack of real currency depreciations and appreciations further aggravated the imbalances.[114] The oil price increases in 1973/1974 were considered temporary, and governments in industrialized countries decided to finance the balance-of-payments deficits rather than to adjust the economy to the new factor prices.[115] Moreover, their fiscal policies attempted to maintain effective demand. However, the expansionary fiscal policies in the OECD countries could not prevent the stagnation of investment. The oil price increase had exposed the pre-existing weakness of many companies' competitive position in the market. In the 1960s expansion, companies had diversified activities beyond their traditional branches and markets, and labour costs were pushed upwards in the booming economies. The location of manufacturing started to change when increasing labour costs and new process and communication technologies made internationalization of production attractive. The new technologies allowed companies to locate the various parts of the value chain in those locations where manufacturing was cheapest. Already in the 1960s, employment in the labour-intensive textile industry in the OECD countries decreased as a result of production moving elsewhere. The oil- price crisis intensified the pressure on the cost structure of companies in the OECD countries, which was further fuelled by inflation, and many were forced to restructure their business. As a result, investment in the OECD countries stagnated, demand for capital decreased and, given the already loose money supply, international inflation increased faster than nominal interest rates. Towards the end of the 1970s, real interest rates were in fact negative.

The combination of inflationary fiscal policies in the industrialized countries with the recycling of petro-dollars through the international private banking system had increased the international money supply when demand for money was relatively low in the traditional markets for this

[114] P.R. Krugman, *Currencies and Crises*, The MIT Press, Cambridge, Mass., 1993, pp. 3-25

[115] M. Garritson de Vries, *op. cit.*, pp.117-21.

capital, i.e., the industrialized countries. However, demand for capital in the developing countries was large. In particular, Asian and Latin American developing countries' demand for capital increased as a result of the balance-of-payment deficits after the oil-price increase and as a result of the recently, more widely adopted, export-oriented industrialization strategies. The petro-dollars started to flow to these countries. Between 1973 and 1988, the debts of developing country increased from $130 billion to $1,162 billion, of which $484 billion was owed to commercial banks and $485 billion to official agencies[116]. Unfortunately, the capital flows were only partly invested in the productive sectors of the economies, and a substantial part was used to finance the balance-of-payment deficits and/or consumption. When in the early 1980s the second oil-price increase[117] compelled the industrialized countries to implement tight monetary policies to fend off inflation, the developing countries were confronted with high interest rates and a strongly appreciated dollar, which made interest payments and the repayment of principal nearly impossible. The debt crisis that broke out in 1982 involved many countries that had borrowed on the international capital markets, and these countries were forced to negotiate structural adjustment programmes with the IMF.[118] A long period of deflationary pressure on their economies seriously impeded their process of development, and subsequently suppressed the growth of the demand for oil.

The debt crisis in many developing countries had a large impact on the role of the state in these countries. The state had taken a leading role in the

[116] B. Eichengreen and P.B. Kenen, *op. cit.* In P.B. Kenen (ed.), *Managing the World Economy, Fifty Years after Bretton Woods*, Institute for International Economics, 1994, p. 41.

[117] D. Yergin, *op. cit.*, 1991, pp. 674-714. In 1979, the Shah of Iran was expelled from his country and a new government took over. In the months of unrest preceding the Shah's removal, oil industry workers went on strike. In the months after he left, the country remained in political chaos, and Iran virtually stopped exporting oil. Japan imported much of its oil needs from Iran, and was forced to replace its Iranian imports by oil bought on the spot market. Since this market was very small, i.e., only 10% of total traded oil went through the spot market, spot prices shot upwards. There was never a physical shortage of oil on the world market because Saudi Arabia managed to fully compensate for the Iranian supplies. However, the fear of shortages so soon after the 1973-1974 crisis, drove prices up.

[118] M. Garritsen de Vries, *op. cit.*, pp. 207-42.

economy, particularly in the industrial sector. The switch, in the early 1970s, from import-substituting development policies to export oriented policies (or in some countries a mixture of the two), had opened up international capital markets for governments and companies. The export-oriented strategy liberalized the economies to some extent, which, given the lack of demand for investment capital in the industrialized countries, made these countries attractive recipients of capital. The relatively low interest rates and the dollar exchange rate in the 1970s had accommodated the large inflow of capital in these countries, and the share of international commercial banks in financing developing countries' external debts increased from 30-35% in the early 1970s to around 50% in the late 1970s.[119] The developing countries preferred the commercial banks over the international financial institutions like the IMF and the World Bank because the private banks attached no special conditionality to their loans with regard to economic policy. The commercial loans were available against flexible interest rates. However, private lending resulted in lower average maturities towards the end of the 1970s.[120] As long as world inflation continued to increase, real interest rates remained low, and banks were prepared to roll over loans frequently, capital was extremely cheap for the developing countries.[121] The immense expansion of international capital markets, coupled with their growing liquidity, stimulated the industrialization in a number of important Latin American and Asian developing countries.[122] Moreover, the internationalization of production continued on its course, increasing the penetration of foreign private capital in some more liberalized developing economies.

[119] K.W. Dam, *The Rules of the Game, Reform and Evolution in the International Monetary System*, The University of Chicago Press, Chicago/London, 1982, pp. 293-6.

[120] G. Bird, *World Finance and Adjustment, An Agenda for Reform*, Macmillan, Basingstoke, 1985, pp. 183-4.

[121] S. Strange, *op. cit.*,1988, Pinter Publishers, London p. 107; C. Edwards, *The Fragmented World, Competing Perspectives on Trade, Money and Crisis*, Methuen, London, 1985, p. 188.

[122] Bank of England Quarterly Bulletin 71, 1990, 'The Growth of the Market for Syndicated Eurodollar Loans'. In H.S. Scott and P.A. Wellons (eds.), *International Finance, Transactions, Policy, and Regulation*, University casebook series, Foundation Press, New York, 5th edition, 1998, pp. 565-78.

4.3 LIBERALIZATION

After the second oil-price increase of 1979-1980, the industrialized countries embarked on tight monetary policies to fight inflation and tried to restrain the growth of budget deficits (except for the United States, which combined a tight monetary policy with public expenditures on 'Star Wars').[123] The money supply tightened and real interest rates increased when inflation declined quicker than nominal interest rates. When, as a result of the American economic policy, the dollar began to appreciate sharply, the developing countries simultaneously faced a situation of sharply increased interest payments and payments on principal in domestic currency terms. The increase in the dollar exchange rate and the interest rate occurred at a time when, on the one hand, their loans came to maturity and, on the other hand, the second oil-price increase had created new balance-of-payment deficits that needed to be financed. For the developing countries that had heavily borrowed on the international capital markets, a period economic contraction had begun. In the ensuing stabilization process, the IMF attempted to slim down the role of the state in the economy.

From 1982, when Mexico announced it could no longer service its interest payments, and loans had to be restructured, the IMF and the World Bank gained importance again.[124] The capital markets shifted their attention away from those developing countries that had become liabilities to the private banking system. The commercial banks discovered painfully that they had not covered their risks sufficiently. Although they had covered their exchange rate risk and interest rate risk by making most of the loans in dollars and tying their interest rates on the London Interbank Offer Rate (LIBOR), the default risk was not covered. The default risk had actually increased by covering the other two risks.[125] Incomplete risk management aggravated the banks' exposure to non-performing loans. The banks had 'secured' their loans with variable interest rates, dollar loans and government guarantees, but this proved insufficient. Moreover, they had created a mismatch between the incoming short-term petro-dollars and the outgoing longer term-loans to developing countries. The banks were in deep trouble when oil-producing countries increased their investments and rolled over

[123] M. Garritsen de Vries, *op. cit.*, pp. 196-203.

[124] S. Strange, *op. cit.*, 1988, pp. 110-1.

[125] B. Eichengreen and P.B. Kenen, *op. cit.*, p. 43.

a much smaller part of their loans to the banks. The 1980s debt crisis was a fact and threatened the international financial system. The IMF had to come to the rescue the countries, and indirectly rescued the highly exposed banks.

Reluctantly, the developing countries turned to the official financial institutions, the IMF and the World Bank, to help them restructure their loans and as well as their economies. The IMF and the World Bank managed to commit some of the larger international commercial banks to underpin their efforts to restructure loans, but they could not prevent that the capital flow reversed in the 1980s[126], and the developed countries became net receivers of capital from the developing countries. Particularly, as a result of high interest rates demand for capital in the United States competed successfully with demand elsewhere in the world, and capital flowed to the United States.

The ability and resources of the IMF and the World Bank to quickly turn around the economies of the debt-ridden developing countries were, however, limited. Due to the debt crisis, commercial loans dried up, and foreign investments (FDI) would be hard to attract, unless a more liberal trade and investment climate was implemented. Particularly, the IMF used its position to demand liberalization of the economy.[127] Measures to break down trade barriers were part and parcel of the restructuring programs and were intended to lower or abolish subsidies on basic needs and to industries, and to allow a larger degree of foreign capital to penetrate their economies, on the one hand, to improve the efficiency of local firms, and on the other hand, to provide capital in the economy without needing to borrow. The IMF opted for the international competition strategy, and although they were pushing for regulatory reforms on the domestic market, they accepted that these would be slow in coming. The 1980s types of reform, however, increased the developing countries' vulnerability to financial crises, as they chose liberalization without introducing regulatory reforms of the domestic economy. Furthermore, vulnerability increased when the IMF allowed

[126] The share of commercial banks in total developing country debt declined relative to the share of official agencies. (cf. B. Eichengreen and P.B. Kenen, *op. cit.*, p. 41).

[127] J.D. Sachs, 'Conditionality, Debt Relief, and the Developing Country Debt Crisis'. In J.Sachs (ed.), *Developing Country Debt and The World Economy*, NBER, University of Chicago Press, 1989, pp. 275-84.

currencies to be pegged to the dollar at a rate that was unsustainable and when they failed to exercise their mandate to recommend an adjustment. When portfolio capital started to flow into emerging market economies in the 1990s, the next generation crisis was in the making. Liberalization, and to some extent, privatization do not work without regulatory reforms in the domestic economy.[128]

In fact, to some extent, this push for liberalization and privatization by the IMF and the World Bank changed the institutional economic structure of many of the developing countries, whose state involvement in the economy had been particularly strong. The state, after many years of playing a leading role in the economy, was stimulated (or even coaxed) to divest itself of this role and to restrict itself to providing a stable macro-economic environment, a role that states in developed countries were also increasingly forced to assume. The debt crisis slowly abated when and the (oil-consuming) developing countries, after a lost decade, started to expand again. The depreciation of the dollar in 1985[129] and the oil-price decline in 1985-1986 were instrumental in this expansion.

The international capital markets continued to grow at a high pace, and when in the early 1990s the industrialized countries went into recession, capital again started to flow to developing countries on a large scale. Emerging markets in particular were the main recipients of these private capital flows. In 1996, net external financing of emerging markets reached $298 billions.[130] In the period 1990-1997, net equity flows to emerging markets increased from approximately $20 billion in 1990 to $133 billion in 1997, of which about two-thirds were Foreign Direct Investments (FDI).[131] In 1996, the net flow from commercial banks to emerging markets was $103.5 billion. The growth of capital flows into emerging market economies came drastically to an end in the period 1997-1999, when the Asian crisis broke out in May 1997, Russia defaulted on its loans in August 1998

[128] See Chapter 3, section 3.4.

[129] B. Eichengreen, *International Monetary Arrangements for the 21st Century*, The Brookings Institution, Washington D.C., 1995, pp. 55-6. In the Plaza Agreement of 1985, the G-5 (US, UK, France, Germany and Japan) agreed to the depreciation of the dollar. They intended to bring about a soft landing, but instead it became a hard one. This dual decline dramatically reduced the oil producers' export income.

[130] Institute of International Finance, *Annual Report 1997*, Economic Analysis, p.1.

[131] *Ibid.*, p. 1-2.

and Brazil, despite a $41.5 billion stabilization package, had to abandon the dollar-real exchange rate policy in January 1999. After the Asian crisis, the IMF switched its focus to regulatory reforms on the domestic markets as a precondition for integration into the world economy.

4.4 DEBT AND OIL-PRODUCING COUNTRIES

In the mid-1980s, the second wave of countries defaulting on their debts to a large extent involved oil-producing countries. The depreciation of the dollar in 1985 and an oil-price collapse in 1985-1986 caused major defaults among oil-producing countries.[132] Particularly, countries that had borrowed heavily against their oil reserves, like Mexico, Venezuela, Nigeria, Algeria, Gabon and others, were deeply affected by the change in monetary policy of the United States. These capital-deficient OPEC member states had suffered from current account deficits since 1981 (except for 1984-85), while the capital surplus member states encountered deficits in the period 1983-84, 1986-88 and after 1991.[133] The combination of a depreciated dollar and low oil prices reduced real oil income to such an extent that both the balance-of-payments and the governments' budgets were in deficit. Moreover, due to capacity constraints most of these oil-producing countries were unable to increase production to any substantial degree. The inability to compensate the price decline with higher export volumes, as Saudi Arabia had been able to do, structurally reduced income from oil.

The painful restructuring process that followed has yet to be completed. The problem for the oil-producing countries was that, as a result of the Dutch Disease[134], the government had to subsidize other industries. The decline in government oil income created a double jeopardy, because the non-oil sectors were unable to survive without government subsidies and trade restrictions. The depth of the recession was aggravated when this type of government support had to be reduced. This was partly due to lower oil income and partly to IMF conditionality. In many oil-producing countries, investment projects had to be cancelled or at least shelved until further

[132] R. Dornbusch, 'Debt Problems and the World Macroeconomy'. In J. Sachs (ed.), *op. cit.*, pp. 299-311.

[133] T.L. Karl, *The Paradox of Plenty, Oil Booms and Petro-States*, University of California Press, Berkeley, 1997, Table A-7, p. 256.

[134] A. Gelb, *op. cit.*, 1988, Chapter 2.

notice. Particularly projects in oil processing were shelved, and the countries seemed to be forced to return to their role of crude oil suppliers.

As a result of the economic perils, the relationship between governments and the national oil companies came under strain.[135] The governments were predominantly focused on the wider economic problems, while the state oil companies were more and more focusing on strategies that would optimize their oil business. It was in this period that the lack of international investments and lack of vertical integration became a major constraint for the state oil companies. The major private oil companies had engaged in FDI to secure supply and the state oil companies wished to do the same to secure demand.

4.5 The First Oil War

The Gulf crisis of 1990-1991 was in many ways a war over oil.[136] Although former President Bush only once referred to oil in a television address early on in the crisis, oil was always prominent on the agenda. The occupation of Kuwait had promoted Iraq to become holder of about 20% of OPEC production and 25% of the world's proven oil reserves, nearly equivalent to Saudi Arabia's[137]. Such a change in the balance of oil power was unacceptable to the United States, the United Kingdom, other European countries and Saudi Arabia. The run-up to the conflict was also dominated by oil issues, when early in 1990 Iraq ran into economic problems and required oil prices to go up. Iraq was unable to raise revenue through the expansion of production and exports and therefore requested production cuts from OPEC member states. It wanted to reduce production to a level where the oil price would become $25 a barrel. Iraq had just agreed to a cease-fire with Iran and had incurred a large debt of around $70 billion to other oil-producing countries (of which $10 billion to Kuwait alone), Russia ($8 billion) and various OECD countries. The oil-producing countries around the Arabian/ Iranian Gulf were not interested in a change of policy, and Iraq began to

[135] See Chapter 7.

[136] During the conflict, Robert Mabro of the Oxford Institute for Energy Studies published a study which he called "The First Oil War", Oxford, 1991.

[137] D. Yergin, *op. cit.*, p. 772.

threaten them.[138] In the summer of 1990, an agreement to cut production was reached and $21 a barrel became the target price. Iraq interpreted the undertaking to guarantee these production cuts for only a few months at a time, particularly on the part of Kuwait, to mean that there would be no certainty that oil income would increase. The soured relationships among the oil-producing countries in the Iranian/Arabian Gulf region prevented any further agreement on this issue, and Iraq decided to push the issue by the use of force against Kuwait, in their eyes an insurgent province of Iraq anyway.[139]

The occupation of Kuwait in August 1990 was countered by a total economic embargo against Iraq, and the build-up of "alliance" troops in Saudi Arabia along the border (Desert Shield). The "alliance" was initially set up to discourage Iraq from threatening and/or seizing the Saudi and Neutral Zone oil fields. The removal of crude exports from both Kuwait and Iraq off the international market as a result of the economic embargo was compensated by increased production levels in Saudi Arabia, Venezuela and other countries. However, the 5% share of Kuwait's export refineries could not easily be compensated for and created temporarily tight product markets, first in Asia and later in Europe and the United States as well. Both crude and product prices temporarily increased, but eased immediately after hostilities started (Operation Desert Storm). The military costs for the United States, the biggest partner in the alliance, were covered by Saudi Arabia from additional profits on oil exports. Other alliance participants, in particular the United Kingdom, were to be compensated by Kuwait after the occupation had ended.

When Operation Desert Storm started, Iraq set fire to many of Kuwait's oil wells. As a result of the damage to its oil fields, Kuwait could not immediately resume oil production and incurred high investment costs. The production costs per barrel increased and Kuwait lost some of its competitive advantage. This added to the pressure on Kuwait to produce more volume in the years that followed in order to recapture some of the investment costs. In reality, this created the undesirable situation for OPEC in which there was always at least one member state which found itself in

[138] C. van der Linde, De olie-achtergrond van de Golfoorlog, *Armex Defensie Magazine*, January 1992.

[139] J. Crystal, *Oil and Politics in the Gulf: Rulers and Merchants in Kuwait and Qatar*, Cambridge University Press, Cambridge, 1990, p. 107.

special political or economic circumstances (IMF stabilization programmes, war damage, political turmoil, and so on). As a result of this, no effective production policy could be implemented without relying heavily on only one or two member states. The national oil policies of these key member states (e.g., Saudi Arabia and Kuwait) were obstructed by the perils of OPEC. The political pressure to carry the burden of stabilization led to several attempts to cut production. However, at the first sign of cheating among the other member states, production agreements were abandoned, renewed and abandoned again. This resulted in a credibility crisis. In the 1990s, OPEC kept sliding further and further into ineffective market control, and the value of OPEC membership for oil-producing countries declined. The decision of two member states (Ecuador and Gabon) to leave the organization, and implement their own, more private sector-oriented, oil policies, was illustrative in this respect.

4.6 Crisis in the 1990s

The decline in price in 1985-1986 did not translate into a return of buoyant growth of the demand for oil in the industrialized countries. Between 1988 and 1998, growth of crude oil consumption in the OECD was on average 1.27% a year, compared to 3.6% for the Asia Pacific region.[140] The diversification away from oil in certain sectors was more structural, and particularly, natural gas gained importance. In Europe, natural gas had become an accepted source of energy for electricity generation in the late 1980s. Until the mid-1980s, the energy policy of the European Community discouraged such use: each year, for instance, the Commission reprimanded the Dutch government for allowing gas to be burned for electricity generation. The increased availability of gas and the maturity of the European gas market, in conjunction with growing environmental concerns about CO_2 emissions, created a change of heart towards gas as a source of energy for electricity generation. The market for natural gas expanded, and oil was replaced by gas in electricity generation in the US and Europe. Nevertheless, primary energy consumption in 1998 was still dominated by oil in most regions (except for the former Soviet Union and Asia Pacific, where natural gas and coal, respectively, are the dominant fuels).[141] Gasolines and

[140] BP Amoco *Review of World Energy* 1999, p. 9.

[141] *Ibid.*, p. 38.

middle distillates make up the bulk of the consumption of oil products, although there are distinct differences among the various regions in the world regarding their share of gasolines, middle distillates and fuel oil. In general, fuel oil is still more important in developing and emerging economies than in developed economies.[142]

World crude oil consumption increased by 8,315 million b/d between 1988 and 1998, to 71,530 million b/d.[143] Consumption increased in all regions, except in the Former Soviet Union, where consumption declined with 4,605 million b/d. About 50% of the growth in consumption in that period was realized in Asia. The influx of capital into Asia spurred demand for oil, and created the only expansive market for oil. The collapse in the early 1990s of the former state-planned economies in Eastern Europe and the former Soviet Union, the economic recessions in the US (however short-lived) and in Europe, and the stagnation of the economy in Japan all made the oil industry heavily dependent on the economic fortunes of East Asia in the 1990s. In 1997, growth in oil demand ended with the outbreak of the Asian financial crisis. World consumption increased only 0.1% from 1997 to 1998, and oil consumption fell 2.7% in Asia.[144]

The continued inability of oil producers in general, and OPEC in particular, to match supply and demand pushed oil prices to very low levels. In the autumn of 1998, the oil price averaged $10-11 per barrel, a major blow to the oil incomes of oil-producing countries. The pressure to restructure their economy was growing, and they were aware that the state had been unable to create sustainable development, and that OPEC no longer had the power to stabilize prices on the international market for a longer period of time. The dramatic development of oil prices in the Autumn of 1998 and early 1999 spurred OPEC and a few non-OPEC oil-producing countries (such as Mexico and Norway) into another attempt to constrain production. The last attempt dated from a year before, but that agreement covered only 1.2 million b/d in cuts from the production level, which had been raised only as recent as December 1997. The expansion of the OPEC production level in December 1997 by 10% was deemed to be the source of the 1998 glut, but the fall in Asian demand and the failure to assess the

[142] *Ibid.*, p. 12.

[143] *Ibid.*, p. 10.

[144] *Ibid.*

depth of that crisis were equally to blame for the price collapse. The boost in output in late 1997 was meant to discourage further expansion of the investments in the Caspian Sea and Central Asia. The combined effect of an output boost and falling demand was disastrous. Saudi Arabia alone lost $14 billion in oil revenues.[145] The Saudi Riyal had to be rescued twice in the space of 7 months as a result of the economic difficulties.[146] Venezuela expected to earn only $9.5 billion in oil export revenues in 1999, down from $18 billion in 1997.[147]

The March 1999 OPEC agreement to cut production by 7% (about 2 million b/d) compared to the March 1999 production levels was a last attempt to stem the hemorrhaging. The political agreement within OPEC to cut production is looking more serious than the agreement of the year before. The fact that both Mexico and Norway support the agreement again, and that Iran and Venezuela have agreed to the cuts under a new political leadership, shores up the agreement. The compliance rate of the March 1999 agreement was much higher than earlier agreements[148], and the short-term interest to boost prices (and revenues) was large. Furthermore, international oil companies have removed around 500,000 b/d of marginal oil from the market as a result of cutbacks in capital expenditure.[149] These measures have led to a substantial reduction of output, and prices shot up in the space of 3 months from $10-11 a barrel to $20-21 a barrel. In fact, prices increased so quick and so dramatic that Saudi Arabia began to worry about over-shooting their target price.

That success of the March 1999 does not mean that the OPEC is back in control of the market. The pressures that have weakened the organization's market power, compared to the 1970s and early 1980s, are still in force and limit the organization's ability to stabilize the market for a longer period of time. However, in the short run the alternative not to cooperate produced an even worse outcome. The success of the agreement lies in the ability of the OPEC to adapt output to demand and stabilize prices at a level that will

[145] J. Richard, 'Can OPEC Make it Stick?', *Barron's*, May 17, 1999, p. 43.

[146] *Financial Times*, 26 February 1999.

[147] *Financial Times*, 10 March 1999.

[148] J. Richard, *op. cit.* , p. 44 .

[149] *Ibid.*, p. 43.

restrict the danger of cheating and does not encourage too much new supply on the market.

The present agreement will come under strain when the higher oil price will invite non-OPEC oil output back into the market, and will allow capital expenditure to expand again. The economies of the oil-producing countries, including Russia, are still in serious trouble, and the pressure to cheat is likely to increase when hard currency reserves are depleted and international financial obligations need to be met.

A successful production agreement has it own pitfalls, particularly for the staunchest supporters of the agreement. As a result of the agreement, Saudi Arabia had to depart from their 'bottom-line' production level of 8 million barrels a day. As soon as oil prices reached a stable level of $18, and investments in the Caspian Sea region recovered, it is to be expected that they will want to return to their original oil policy and that they will propose a more relaxed production level to maintain prices within their desired range. Rather than wait for cheating to commence, these countries should attempt a soft landing of the oil price. However, soft landings usually end with a bump, because they fail to properly signal the market of their intents.

Such a relaxation is dependent on the ongoing talks between Iraq and the UN about compliance with the UN resolutions. A certain relaxation of the restrictions on Iraq's oil industry appears to be under way. Already, Iraq is producing more than strictly allowed under the UN resolution. This is due to the fact that Iraq's production level is determined by a revenue level. The low oil prices of 1998-1999 allowed Iraq to produce a large volume of oil. The present high oil prices should have resulted in a reduction of output, but it did not. A rehabilitation of Iraq will immediately increase oil output available to the world market. Moreover, international investors are ready to commence investments in Iraq's oil industry. The March 1999 agreement will be tested for its robustness when Iraq returns to the international oil market.

Chapter 5

The Failure of OPEC to Secure Economic Rents

5.1 Policy Tool

Since 1973, OPEC has been markedly less successful in reconciling the individual and collective interests of the member states. Differences in oil reserves, production capacity, dependence on oil income and the degree of vertical integration have brought conflicting interests to light concerning price and production levels. At the same time, the international oil market has been undergoing major changes. As a result of these internal difficulties and oil market developments, OPEC has become increasingly unable to secure the lion's share of the economic rents for its member states. The organization was further strained when the weaker OPEC countries experienced difficulties in financing their new rounds of investment in the oil industry. They were increasingly looking for foreign investments to fill the financing gap. However, attracting foreign investment becomes easier if agreements within OPEC to limit production are abandoned. However, as a result of more private investment, the downside of reducing state control over the industry in the member states is that the position of OPEC to regulate the market will be further weakened.

For the member states OPEC was one of the prime mechanisms to realize economic rents from oil, and subsequently, to underpin their efforts to finance economic development. This principle was laid down in the statute of the organization. However, its initial success in the 1970s created inflated expectations of the organization and ignored the underlying diverging short-term and long-term economic interests of the member states. Shortly after the oil-price increase of 1979-1980, OPEC's influence on the international oil market started to weaken, and OPEC was forced in a position of swing producer in a contracting market. The member states had taken too little notice of the changing international oil market, and short-term gains were preferred over long-term ones. The agreements on output restrictions have been short-lived, and had a fairly low compliance rate. OPEC's record of market stabilization has not been good. Moreover, OPEC failed to produce stable income from oil for the member states and, as a result, economic difficulties have intensified.

5.2 SECURITY OF DEMAND

The international oil markets have been struggling with periodic surpluses for more than 15 years. The relatively high oil prices of the early 1980s encouraged substitution, energy saving and more efficient use, which have made the demand for oil more elastic. A particularly important aspect in this respect was that oil consumption per unit of GNP dropped substantially in the industrialized countries, as a result of which demand, after the drop in oil prices in 1986, increased much less than OPEC had hoped. Only demand from the Southeast Asian countries increased significantly. To make matters worse, the recession in the USA and the UK in the early 1990s and the reduced growth in the other important consumer countries exacerbated OPEC's sales problems, as a result of which the conflict between the member states over policy issues was intensified.

The development of mineral energy consumption is strongly linked to the economic development of countries and particularly the level of industrialization they have achieved. Although, coal was the dominant energy input in the economy during the earlier stages of industrial development, it was increasingly substituted by oil (although not at the same time and rate in each country), when industrial production techniques advanced and interfuel competition favoured oil consumption. The era of mass industrial production and consumption had corresponded with a sharp increase in oil consumption in the industrialized countries. The share of oil in the energy mix fist increased in the United States, followed by Western Europe and Japan after 1945. Oil consumption in the OECD countries had increased unabated until the oil price increase of 1973-1974. Only after this oil price increase did the share of oil in the energy mix decline and later stabilize, when both coal and gas substituted oil. Nevertheless, the share of the OECD countries in world energy consumption remains high. In 1998, they consumed around 58% of the total world primary energy consumption, 63% of world oil consumption, 55% of world gas consumption, and 48% of world coal consumption.[150] The share of emerging market economies in world primary energy consumption did improve, but it is still modest. Their share (excluding Central Europe and the former Soviet Union) was 30% in

[150] BP Amoco *Statistical Review of World Energy* 1999, p. 35.

1998. This large group of countries consumed 31% of world oil consumption, 20% of world gas production, and 44% of world coal consumption.[151]

The price increases in 1973-1974, but especially those of 1979-1980 invoked a change in the growth of demand for crude oil and oil products. The oil price increases in 1979-1980 fed the already high world inflation rates, and consumer governments, also facing substantial budget deficits, started to deflate their economies. In the early 1980s, tight fiscal and monetary policies caused a worldwide economic downturn, which led to in a decline in oil consumption. As a result of more expensive oil, structural changes in demand were achieved by more efficient use. The gains in consumption efficiency were accompanied by OECD governments' energy policies, geared to decreasing import dependency, geographical diversification, and changing the energy mix away from oil. For instance, electricity was increasingly generated by coal, nuclear power, and later also gas. The expansion of electricity consumption gave consumer governments more control over the energy mix in their economies.

The growth potential of the demand for oil in developing countries remained unabatedly large. However, the oil price increases in the 1970s and the response to the subsequent lack of international economic equilibrium stifled the expansion of their economies and subsequently also the growth rate of oil consumption. Despite the severe economic crisis in the developing countries, oil consumption continued to grow, but at a lower rate.[152]

In the early 1990s, OPEC has been producing around 24-25 million barrels per day, which amounted to approximately 38% of world production. Of this, Saudi Arabia was responsible for around 7.5 to 8 million barrels per day. The scope for the remaining member states came under pressure from the Saudi share (which had increased since the Gulf War), the continuing threat of Iraq's return to the international oil market and the resumption of production in Kuwait. For the 4th quarter of 1992, the OPEC ministers therefore decided against setting a real ceiling for OPEC production. Instead, OPEC was to maintain a market share of 24.2 million barrels, and quotas were not divided. This type of loose agreement did not work, because the member states were very soon urged to curb production

[151] *Ibid.*, p. 35.

[152] *Ibid.*

again, when monthly figures showed that production had slowly increased above the agreed level.

On 18 September 1992, at the end of the meeting, Ecuador announced its resignation from OPEC as a full member. This announcement represented the open admission that, in fact, over the past few years, the interests of the individual members had not been served by the collective action OPEC had undertaken on the oil market. The cartel theory and insights granted by game theory predict such dissolution of cooperating groups of market participants as the conditions of cooperation change or disappear over time.[153]

In December 1997, OPEC made the remarkable decision to increase the production shares of the member states. In the midst of the unfolding of the Asian Financial crisis, this decision was inspired by conflicting interests. Some member states wanted to expand short-term oil revenues, while Saudi Arabia wanted to push the price under the $20 a barrel level to prevent the expansion of investments in the Caspian and Central Asian area.[154] The expansion of output combined with a severe drop in demand as a result of the economic crisis in Asia created a price fall that was very hard to bring under control again. Only in March 1999, after an agreement in 1998 had failed to take effect due to persistent cheating of some member states, a new agreement to curb output at 25,676 thousand barrels a day managed to hold and prices recovered very quickly to $20-21 a barrel. Saudi Arabia took an 8% cut in its production level (to 7,438 thousand b/d) against a production capacity of 10,800 thousand b/d.

5.3 The aims of OPEC

Market conditions have changed so much over the years that OPEC, an intergovernmental organization, is no longer able to realize its main objectives. These objectives stress in particular the importance of stable prices and income from oil export for the member states.[155] The pursuit of

[153] See H.W. de Jong, *Dynamische Markttheorie*, Bedrijfseconomische Monografieën, Stenfert Kroese, Leiden/Antwerpen, 1985, p. 151 f.f.

[154] J. Richard, *op. cit.*, p. 43.

[155] In Article 2 sub A, B, and C, Chapter 1 of the Statute of the Organization of the Petroleum Exporting Countries (OPEC), the main aims are set forth: the coordination and unification of the petroleum policies of members states; the determination of the best means for safeguarding the individual and collective interests

a coordinated or even uniform oil policy by the member states was considered of highly important for the realization of these objectives, so that both collective and individual objectives could be pursued as fully as possible. However, this requires converging developments in the oil industries and economies of the member states.

As long as the member states lacked influence on the price and production levels (international oil companies had oil concessions), the member states had a common objective: obtaining influence over price and production. In the market conditions of the late 1960s and early 1970s, increased influence for the member states automatically led to higher oil incomes. At that time, the heterogeneousness of the member states played no significant role.

In game theory terms, it can be asserted that before and during the oil crisis of 1973 there were sufficient common interests to enable agreement to be reached. Market conditions at the time allowed both short-term windfall profits and control over the levels of production to be achieved. Thus, the members were involved in a cooperative game without conflict, and they could reach a Nash equilibrium.[156] Stated differently, it could be said that the international oil market in 1973 satisfied the criteria for the formation of a strong cartel. Demand was price inelastic and supply was strictly concentrated. As a result, after the Tripoli and Teheran agreements of 1971, in which the member states gained more say in prices and production, supply was also inelastic in the short term. The OPEC members produced around 55% of world production in 1973 and accounted for around a 73% share of oil exports.[157] In that respect, the year 1973 was a

of the member states; stability of prices in international crude oil markets in order to eliminate harmful and unnecessary fluctuations; due regard to the interests of the producing nations and to the necessity to secure a steady income; an efficient, economic and regular supply of petroleum to consumer countries; and a fair return on capital to those investing in the oil industry. OPEC, *Official Resolutions and Press Releases*, 1960-1990, The Secretariat OPEC, Vienna, p. 32.

[156] A Nash equilibrium is an equilibrium "where no player has an incentive to deviate from his strategy given that the other players do not deviate". E. Rasmusen, *Games and Information, An Introduction to Game Theory*, Basil Blackwell, Oxford, 1989. p. 33.

[157] OPEC, *Facts & Figures, a graphical analysis of world energy up to 1985*, 1986, p. 25.

unique year for the OPEC members, and one which will never be repeated. From then on, centrifugal forces on OPEC increased.

The control over oil production was completed in the period 1974-78, when the interests of the international oil companies were nationalized and organized in national oil companies. In addition to the differences in oil reserves and production capacity, the creation of national oil companies introduced another latent source of conflict within OPEC. In fact a wide range of conflicting interests within OPEC increased because: (a) the interests of the state and the interests of the national oil company are not necessarily the same and both alter over time; (b) the market strategies of the various national oil companies increasingly varied; and (c) differences in the degree of internationalization and vertical integration of the national oil companies impose diverging demands on the price and production policy of the member states, and consequently OPEC.

Given the above, it is conceivable that not only will there be more conflicts between the member states, but also between the member states and their national oil companies. Such a conflict has already emerged in Venezuela, where the state oil company PDVSA complained about the subordination of company interests to the resolution of conflicts in the OPEC context. The company's management was annoyed with the oil minister, who had approved production limitations to settle (political) disputes between Arab member states. In addition, there have been tensions concerning the internationalization and investment policy of PDVSA, which, according to the government, was in conflict with government policy.

5.4 Heterogeneousness

The year 1973 was both a year of great success and the year that marked the start of fundamental conflicts of interest among the member states. These conflicts made the formulation of a common policy that would embrace the individual interests of all members impossible.[158] The persistent differences between the member states in the size of their reserves, their production capacity, their economy's absorption capacity, their dependence on oil revenues, their degree of vertical integration and the degree of internationalization of their national oil companies resulted, both in the short and the

[158] For a discussion of an asymmetric model of the leader-follower type see A. Jacquemin, *The New Industrial Organization, Market Forces and Strategic Behaviour*, Clarendon Press, Oxford, 1987, pp. 72-4.

long term, in contradictory interests between member states concerning the desired price and production levels.[159] In game theoretic terms, after 1973, the OPEC countries were confronted with a prisoners' dilemma -a non-cooperative game with conflicts. This game theoretic model presumes that the member states maximize their own utility functions and can make no credible commitments; although the members can agree to cooperate, they will act in such a way as to maximize their own income if the chance allows.[160]

The inability on the part of the member states to make binding agreements is inherent in the intergovernmental character of the organization. Above all, this intergovernmental character, and the weight that it accorded to the member states' sovereignty over their own natural resources, prevents the imposition of sanctions to ensure that the policies are implemented. Within OPEC, the members retain sovereignty over oil policy. Thus, in the 1980s, individual OPEC member states not only had large financial incentives to cheat, but also lots of freedom to do so.[161] Moreover, John Elster emphasized that in a heterogenous group, collective action requires a great deal of negotiation as to the distribution of advantages and disadvantages. In his view, the inability to agree on distribution is a far greater threat to the stability of the cartel than the free rider problem.[162]

As the international market for OPEC oil began to shrink at the beginning of the 1980s, the distribution problem increased in gravity, and conflicts intensified. The conflicts operated on OPEC as centrifugal forces and made that collective action increasingly difficult. In some member states, the gap between the goals of their own national oil policy and the agreements in OPEC became increasingly wider. In addition, the regional political conflicts in the Gulf increased in severity, and further complicated economic disputes.

[159] C. van der Linde, *op. cit.*, 1991, Chapter 5.

[160] E. Rasmusen, *op. cit.*, p. 29.

[161] T.H. Moran, *Oil Prices and the Future of OPEC. The political economy of tension and stability in the Organization of Petroleum Exporting Countries*, Research Paper R-8, Resources for the Future, Washington D.C., 1978, p. iii.

[162] J. Elster, *The Cement of Society. A Study of Social Order*, Cambridge University Press, Cambridge, 1989, p. 174.

5.5 POLITICAL AND ECONOMIC DYNAMICS

After the oil price crisis of 1979-80, the tensions about which price and production policy to pursue became increasingly obvious. Members with a relatively low reserve and production capacity, such as Algeria, Ecuador, Gabon, Nigeria and Indonesia (supported by Iran and Iraq which were engrossed in an expensive war) were proponents of a high price supported by production restrictions. These countries operated in accordance with their own rational interest. The present value of a barrel of oil was higher for a country with smaller proven reserves than for a country with large proven reserves. The price level preferred by these countries in 1980 and 1981 was largely determined by their income requirements and was supported by the short-term price inelasticity of oil.

Countries like Kuwait, the United Arab Emirates (UAE), and Saudi Arabia have an interest in a relatively low price. Defending a relatively high oil price by means of production restrictions curtails demand and stimulates replacement of oil by other sources of energy. The rational economic interest of these countries was subordinated to strategic political interests that resulted from the war between Iran and Iraq and the economic interests of the other member states. The two warring countries, given their lower exporting capacity, could only finance the war by means of export of oil at a high price. Kuwait, Saudi Arabia, and the UAE could not withstand the political pressure imposed by Iran and Iraq, which knew that the high oil price policy had the support of the other member states. Moreover, in 1980-1981 Saudi Arabia lacked the surplus capacity to force the oil price down to more acceptable levels.

5.6 TRANSFER OF INCOME

The stabilization of prices at a high level meant extensive restrictions on production for Kuwait, the UAE, and Saudi Arabia. Between 1981 and 1986, these countries produced 6.6 million barrels per day less, of which the lion's share of the cutbacks fell to Saudi Arabia. The proponents of a high oil price policy also had to limit production, as the market for OPEC oil was shrinking (from 23.4 million barrels per day in 1981 to 17.3 million barrels per day in 1985), but this was by no means to the same extent as for Kuwait, the UAE and Saudi Arabia.[163] The OPEC policy of the early 1980s therefore involved a large redistribution of oil income from the economi-

[163] C. van der Linde, *op. cit.*, p. 175.

cally weaker members. The absorption of the costs by a small group of member states allowed the other member states to abuse the system.

In the short term, such a policy of production limitation favours states with relatively low oil reserves and production capacity. However, the problem of the distribution of costs of the agreement increased over time, particularly when the political and economic strategic interests of the three states in question were no longer served by a high oil price, but rather required the maximization of oil revenues. In December 1985, Saudi Arabia announced that it no longer supported such a policy, not merely because it has lost a great deal of its market share, but more particularly because it had itself encountered financial problems. The value of OPEC oil exports had already dropped from its all-time high in 1980 of $284 billion to $131 billion in 1985.[164] With its rejection of the production- limiting policy, Saudi Arabia also resigned from its position as swing producer within OPEC. Together with Kuwait and the UAE, Saudi Arabia realized a completely different market strategy. The increased volume of spare production capacity in these countries allowed them to push the price of oil down. The income advantages of a lower price and a larger oil export volume could be enjoyed by these three countries in particular. However, this new course led to a reduction in oil revenues for the smaller OPEC producers such as Ecuador, which had less spare production capacity; the disadvantages of the liberalization of the oil price were largely experienced by them.[165]

5.7 Increase in Conflicts

After the steep price drop of 1986 it took some time for the members states to attempt to stabilize the oil price again. By then, the smaller exporters had already had to sacrifice oil revenues, and they had to accept an oil price of $18 dollars per barrel, which better suited the long-term interests of the larger OPEC producers. The conflicts over which price and production policy to pursue also played an important role in the conflict between

[164] OPEC *Annual Statistical Bulletin* 1991, p. 6.

[165] The value of oil exports of the OPEC member states dropped to $77 billion in 1986, increased to $94 billion in 1987, and dropped again to $87 billion in 1988. OPEC *Annual Statistical Bulletin* 1991, p. 6.

Kuwait and Iraq, which led to the Gulf War in 1990.[166] From 1986 to January 1990, Iraq mainly supported the low-price policy of the other three member states, because at that moment in time, it believed Iran would suffer more from low prices than itself. Additionally, Iraq received support from the other Gulf States. However, when it emerged that this support evaporated after the cease-fire with Iran, and that Iraqi oil production could not be increased, Iraq abandoned its support for this policy. Iraq then started to plead for stricter adherence to the production- limiting policy by the member states so that the oil price could rise to $25 per barrel. The difficult negotiations over the Iraqi proposal in the spring of 1990 led, in July 1990, to a compromise price objective of $21 per barrel and an interim agreement on production restrictions, both of which indicated the reluctance to change policy.

After the Gulf War, Saudi Arabia again increased its market share from the 25% claimed in 1986 to 34.5% in 1991. It therefore appeared that the power struggle within OPEC had been resolved in favour of Saudi Arabia and the other surplus-producing states. In fact, the abandonment of the price policy in 1985 amounted to the recognition of the inherent conflicts between the member states that had been mounting since 1973.

5.8 Under Investment

The long period of relatively low oil prices (in part due to the depreciation of the dollar in the same period) and the limited growth of exports, put oil incomes under pressure for most OPEC member states, particularly those which were unable to raise their volume of exports due to problems with capacity. The OPEC member states are all largely dependent on revenue from the export of oil. In countries like Algeria, Nigeria, Ecuador, Venezuela, Gabon, Indonesia, Iran and Iraq, this dependence is greater in the sense that the oil industry has to compete with other sectors for the investment of oil revenues, while at the same time the incomes are highly significant for governments' budgets. These countries have also built up considerable debts, and in a number of countries the economic situation has resulted in large social tensions. Investment in exploration and development was depressed for quite some time, and many countries saw the degree of capacity utilization increase in the period 1989-1992.

[166] C. van der Linde, 'De olie-achtergrond van de Golfoorlog', *Armex Defensie Magazine*, January 1992.

Particularly from 1986 onwards, investment by non-OPEC member states has been low as a result of the buyers' market, leading to under- investment in the early 1990s. In the early 1990s, OPEC calculated that the growth of demand, in combination with the reduction of non-OPEC production, would lead to an extra call on OPEC oil of 7 million barrels of oil per day by the end of the decade.[167] However, the organization was also aware that sustainable capacity was below the capacity of 31 million barrels per day required in the year 2000. To increase capacity to this level, investments totalling approximately $80 thousand million were therefore estimated to be necessary, while for investment in non-OPEC countries another $170 thousand million was needed.[168] The predictions on the additional call on OPEC oil have been proven wrong. In 1999, OPEC's estimated capacity had increased to 32, 970 thousand b/d, while production was capped at 25, 676 thousand b/d in order to support the oil price.[169]

Increasingly, because of the precarious state of their public finance, the oil-producing countries are looking to international oil companies to finance new investments. However, many international oil companies have also been struggling to adapt their business operations, because the relatively low oil prices have also affected their cash flow. In addition, environmental protection measures (e.g., cleaner products, processing and the proposed energy/carbon taxes) have led to greater investment demands on the production, processing and distribution of oil products, and have complicated the prediction of developments in demand.

5.9 The Environment and Investment

The investment opportunities for private investors in crude oil exploration and production have definitely expanded, also in the OPEC countries. One could say that there is a certain amount of competition for private investment capital among the weaker exporting countries. These countries are not only competing for this capital with one another, but also with the investment needs of other parts of the production chain, and with investments needs in newly emerging (oil) markets. Some of these investment

[167] Dr. Subruto, 'How OPEC perceives the energy outlook for this decade', *OPEC Bulletin*, April 1992, p. 12.

[168] *Ibid.*, p. 12.

[169] J. Richard, *op. cit.*, p. 43.

needs are the result of the environmental standards that are imposed on the processing industry by OECD governments. These environmental directives can limit the scope for investment by private undertakings in exploration and exploitation, which in turn delays the development of the new generation of oil production in the OPEC countries. Considering the dependence of the OPEC economies on oil revenues, such a delay could jeopardize future growth. For example, in the same period in which a significant extension of production capacity is required of crude oil, the investment needs of the processing industry are also increasing.

In the 1980s, the refining capacity in the OECD countries was reduced by approximately 10 million barrels per day, while it increased by merely 4.5 million barrels per day in the developing countries. In the 1990s, the refining capacity in the OECD countries increased again, by approximately 3 million b/d to 42.8 million b/d, while capacity utilization slightly increased from 86.3% to 93.1%.[170] The US accounted for most of this growth in OECD capacity. Moreover, refining capacity in the formerly centrally planned economies of Eastern Europe and the former Soviet Union requires urgent rationalization, as a result of which global investment requirements in this sector will only increase in the near future. In Europe, there was a slight increase in capacity in Western Europe, but the decline of refining capacity in Eastern Europe cancelled out growth for the region. Refining capacity in the former Soviet Union declined by nearly 2 million b/d from 1990 to 1998. Refinery throughput shows the disastrous developments in the industry in the former Soviet Union with a decline of nearly 5 million b/d in the period 1990-1998. In 1998 capacity utilization was no more than 44.8%.[171]

Not only was expansion of processing capacity necessary to meet future demand, but existing capacity required upgrading to satisfy demand for lighter oil products and to satisfy newly introduced environmental legislation such as the American Clean Air Act Amendments 1990 (CAA 1990). The capital investments in the oil-processing industry for the 1990s alone were estimated to amount to approximately $200-300 billion.[172]

[170] BP Amoco *Statistical Review of World Energy* 1999, pp. 16-7.

[171] *Ibid.*, pp. 16-7.

[172] A. Seymour, *Refining and reformulation: the challenge of green motor fuels*, Oxford Institute for Energy Studies, Oxford, 1992, p. 2.

The costs of replacing and modernizing the world oil tanker fleet came in addition to this estimate. Oil tankers have also become subject to stringent environmental and health and safety legislation. For example, insurance costs for oil loads will increase as countries continue to apply an unlimited "polluter pays" principle. The Exxon Valdez disaster in Alaska has demonstrated that the costs of a spillage can be enormous. With the reduction of oil production in the OECD countries, increasing volumes of oil are produced away from the consuming markets and transported by sea. Moreover, more oil will be produced by off-shore fields. Subsequently, the risks for the industry increase, which will be translated into higher costs. Depending on the market segment in which the final oil products are sold, these costs may or may not be fully recouped.

5.10 The future of OPEC

Despite the 76% share in the world's known oil reserves, the market power of OPEC member states has become rather weak. Moreover, the short-term capital squeeze seems to exert a strong disintegrating force on the organization. In the coming years, the enormous investment efforts which must be made in all sectors of the industry throughout the world will result in the reorganization of the oil industry. As happened in the past, old structures will disintegrate to be replaced by the new.[173] In the 1970s, nationalization resulted in the replacement of the private sector with the public sector in OPEC countries. Oil interests were brought under the control of national oil companies and the governments received the economic rents and profits directly or in taxes. A few national oil companies, such as KPC (Kuwait), Saudi Aramco (Saudi Arabia) and PDVSA (Venezuela), underwent processes of vertical integration and internationalization, with the objective of guaranteeing markets and long- term economic rents.

The majority of the OPEC countries, however, as a result of outstanding foreign debt, are unable to generate the necessary investment capital to expand their production capacity. These countries are increasingly searching for ways to attract private investment capital. The governments and/or the national oil companies cannot finance these operations themselves, nor do they seem to be capable of facing the necessary management, geological, technical and price risks, and, subsequently, they are looking for

[173] C. van der Linde, *op. cit.*, 1991, p. 203 ff.

partners who will take these risks.[174] Increasingly, these investment risks, or part of these risks, are discarded in exchange for equity oil, whereby the oil reserves are partially privatized. However, for a number of OPEC member states, exercising their sovereignty over property rights through state exploitation of their natural resources is of immense political importance. This forms an obstruction to liberalization and privatization of the oil sector, and not to mention foreign direct investments. For some member states competition among sectors to obtain investment capital has already led to such privatization.

For private investors, however, OPEC membership implies a risk of production restraints enforced by the member states governments' agreement in the OPEC conference. This could impede foreign investment unless the host government is prepared to guarantee production levels. However, providing such guarantees to private investors limits the room for manoeuvre of oil ministers to make arrangements at the OPEC level. Under the ongoing condition of a buyers' market, the members states, if they wish to attract private investors, must either leave OPEC, as Ecuador did, or adapt OPEC strategy. The OPEC strategy should, more than in the past, take the member states' diversity and the new national oil regimes into account. The implication is that the member states can no longer strive for unified price and production policies, and that OPEC's traditional role in the generation of economic rents for its member states belongs to the past. In short, even though the international strategy of the member states brought short-term successes, it failed in the longer run.

[174] T. Wälde, 'Investment Policies in the International Petroleum Industry; Responses to the current crisis'. In N. Beredjick and T. Wälde (eds.), *Petroleum Investment Policies in Developing Countries*, Graham & Trotman, London, 1988, p. 12.

CHAPTER 6

TURNING BLACK GOLD INTO DEVELOPMENT

6.1 WINNERS AND LOSERS

The two oil shocks, followed by the interest rate and dollar exchange rate 'shocks', had, as explained earlier, severe economic effects on oil- consuming developing countries in the early 1980s. Only when oil prices declined and when the dollar exchange rate and interest rates started to fall in the second half of the 1980s did the debt crisis subside somewhat and real economic growth return to these countries. In the oil-exporting countries, who were the apparent beneficiaries of the oil boom economic developments followed a contrary path. In the 1970s, the terms of trade for these countries improved dramatically, although some of the benefits of higher oil prices were inflated away. Unlike the oil-importing developing countries, the oil producers were deemed to be in a unique position to improve their economic performance. It was widely believed that the oil windfall provided them with the means to make a 'big leap forward', and would lead to investments in growth and sustainable development. The results, however, have been disappointing.

6.2 A LEAP FORWARD

The economic development experience of many developing countries in the 1950s and 1960s, when the majority of countries pursued an inward-looking industrialization strategy, had not been very positive. Attempts by countries pursuing this strategy to counter the structural decline of their balance of trade by replacing consumer good imports by capital good imports were unsuccessful. Meantime, export income, which was required to finance these imports, continued to be earned by the traditional primary sectors, and countries increasingly experienced problems in financing their industrialization strategy. In general, growth in developing countries was held back due to a lack of domestic savings, foreign exchange and fiscal revenues, which stunted investment and capital formation.

To overcome these constraints, many countries switched to a more outward-looking strategy in the early 1970s. This involved industrialization, with the resulting industries producing both for the domestic and for foreign markets. The upside of this strategy was that it allowed for economies of scale and, through exposure to international competition more efficient production. This contrasted strongly with the earlier inward-looking strategy, which had implied a high level of protection of the domestic economy. The outward-looking strategy also paved the way for foreign investment capital as a way of overcoming the lack of domestic investment potential. In many cases, the comparative advantage that these developing countries had to offer was a large and abundant supply of labour. The introduction of international competition, however, did not lead to the removal of all restrictions in the domestic economy. Until the outbreak of the Asian financial crisis, a liberal trade policy with limited domestic competition was, with local variations in the level of openness, the dominant development strategy. In the 1980s, the lack of domestic competition was not favoured by the IMF or the World Bank, but these institutions tried to take liberalization as far as they could, politically. After the outbreak of the Asian financial crisis, the Bretton Woods institutions no longer considered domestic competition to be outside their area of competence, but, on the contrary, a vital element in stabilizing economies. The liberalization of a country's domestic economy and balance of payments, and the proper supervision and regulation of its financial sector became part of the new policy standard.[175] In the aftermath, the state has been blamed for much of the effects of the financial crisis. However, Higgott notes that "(...) It was not the involvement of the state in the economy per se that was the source of the crisis rather than that it was the nature of the involvement of the state that was to blame".[176] The same applies to the state involvement in oil-producing countries. Many developing countries, including the oil-producing countries, suffer from insufficient state capacity to guide and coordinate economic change.[177]

In this light, it is understandable that the oil windfalls were seen, at the time, as a great opportunity for the oil-producing countries to parachute themselves into the community of the industrialized countries. The oil-price increase of 1973-1974, combined with the nationalization of their

[175] C. van der Linde, 'Hervorming van het IMF', *Internationale Spectator*, jaargang 52, no. 11, November 1998, p. 580.

[176] R. Higgott, *Review of Globalisation*, paper prepared for The Economic and Social Research Council, June 5, 1998, p. 21.

[177] *Ibid.*

natural resources, provided them with the means to finance the modernization of their economies and develop them to full potential. The oil-price increase in 1973-1974 freed the oil-producing countries from the earlier constraints on growth, except that in some countries (specifically Saudi Arabia, Kuwait, and the United Arab Emirates), the constraint on the labour market was aggravated rather than eased as a result of the windfall. However, for the majority of the OPEC countries, the availability of labour did not, except in terms of technical skills, constitute any major constraint on growth.

The windfall profits were thus initially perceived as an unique opportunity to convert the new oil wealth into sustained economic growth. That is, based on the dominant ideas at that time about growth, development and the role of the state, the countries were thought to be ideally placed to achieve a leap forward in economic development. It is in this spirit that Auty observes that: "The oil shocks of 1973-4 and 1979 each transferred the equivalent of 2 per cent of global GDP to the oil-exporting countries. The resulting windfalls provided unprecedented opportunities for the oil-exporting countries to accelerate economic growth, promote healthy structural change, and speed spatial decentralization. Resource-based industrialization (RBI), which is the further processing of natural resources such as minerals into metals and timber into wood products, played a central role in such efforts."[178] However, twenty years after the first oil windfall, one wonders whether the fate of some countries' 'leap forward' has become similar to Mao's in the 1960s: of the original 13 OPEC member states, seven are experiencing severe economic adjustment problems, and some have political problems to match. Accordingly, Auty also observed:"The high expectations for RBI were largely unfulfilled, raising important questions about the cause of its failure; its wisdom as a development strategy; and the consequences for both the oil-exporting countries and key-sector in global manufacturing industry."[179]

In light of the growing literature on the disappointing economic performance of developing countries favourably endowed with natural resources, i.e., suffering from the resource curse, it may be prudent to take a closer look at the effect of windfall profits' on the economy and at the strategies

[178] R.M. Auty, *Resource-Based Industrialization, Sowing the Oil in Eight Developing Countries*, Clarendon Press, Oxford, 1990, p. 3.

[179] *Ibid.*

of those oil-producing countries in which the governments took an significant leading role.

6.3 The Resource Curse Thesis

The oil industry is an industry in which, typically, large economic rents can be earned because the market price is well above the price required to keep the factor of production into employment and is above the price required to earn economic profits.[180] This encourages firms to compete to capture these rents, and it induces governments to tax it away. However, it is unclear exactly what proportion of economic profits are economic rents, particularly in a highly unstable market with large price fluctuations. Moreover, economic profits may rise above the level of interest rates as a result of monopoly profits, income from risk-bearing, and rewards for innovation, and the level of taxation cannot be easily be determined. Baumol and Blinder state that: "So profits in excess of the market rate of interest can be considered as the return on entrepreneurial talent. But this is not really helpful, since no one can say what exactly what entrepreneurial talent is. Certainly we cannot measure it; (...). Therefore, we do not know how the observed profit rate relates to the minimum reward necessary to attract entrepreneurial talent into the market -a relationship that is crucial for the contentious issue of profits taxation."[181] If profit rates in the oil industry are well above this minimum rate, they must contain a large component of economic rent. In this case, taxing away the excess profits would have no impact on oil production. However, if the profits do not contain much economic rent, taxation could curtail exploration and the production of oil. Gelb underwrites the notion that, since 1973, the oil price in OPEC countries has contained a large portion of economic rent.[182]

The sellers' market in the 1970s placed OPEC countries firmly in command of the crude oil market. The major international oil companies' oil

[180] "Economic rent is said to be earned whenever a factor of production receives a reward that exceeds the minimum amount necessary to keep the factor in its present employment." (p. 753) What makes economic rent so important is that it can "be taxed away without reducing the quantity of input supplied". (p. 755) W.J. Baumol and A.S. Blinder, *Economics, Principles and Policy*, 5th edition, Harcourt Brace Jovanovich, Publishers, San Diego/New York, 1991.

[181] *Ibid.*, p. 755.

[182] A. Gelb, *op. cit.*, p. 7.

reserves and production facilities in the OPEC countries had by and large, been nationalized, and were operated by state oil companies (sometimes newly formed). Producer governments' income from oil either accrued from the heavily taxed revenue of state oil companies or else the revenues went directly to the government and left the companies with operational monies only. In fact, oil revenue was an important, and in some cases the main, source of revenue for governments' budgets. The economies of OPEC countries were all developing economies and had been in need of capital to improve economic and social infrastructures, to increase production capacity, and to diversify the existing production base. That constraint on development had been lifted.

However, the favourable redistribution of income to oil producers posed the problem of how to turn growth potential into real and sustained growth. Most producer countries had only small modern private sectors of the economy, and labour was predominantly occupied in agriculture, handicrafts, trade and services typical of developing countries. In general, the gains in oil revenue were devoted to infrastructure and to diversifying the economy. This meant, on the one hand, strengthening the non-oil sectors, and, on the other, increasing domestic processing capacity in the oil sector in order to reduce the dependence on crude oil exports. OPEC countries also extended state control over the oil sector by nationalizing the local assets of the international oil companies in their countries. When governments started to spend their oil revenues, the fact that the private sector was relatively small and that the institutional and social change that was needed to facilitate the absorption of the oil revenues by the private sector would only take place gradually, also served to increase the governmental control over other sectors of the economy.

Since oil revenue accrued predominantly to governments, governments took upon themselves the task of investing this revenue in the economy in such a way that the economy would become more diversified and be set on the path towards sustainable non-oi- industry related growth. Moreover, the private sectors in the oil-producing countries were very small and could not absorb the windfalls. However, there are a number of drawbacks that result when the state takes the leading role in the economy, because it is particularly difficult to create the proper spill-over effects from the boom sector to other sectors of the economy without distorting the resource allocation of the country. This is often referred to as 'the Dutch Disease'.[183] Auty states that: "Developing country governments find commodity windfalls difficult to employ prudently. They tend to promote too rapid a rate of domestic absorption so that they amplify the boom effects. This in turn expands the service sector and corrodes the attraction for both investors and employees of the lagging (non-mining tradeable) sector. Such amplification of boom conditions is also fed through the demonstration effect of high mining wages which feeds popular pressure for the immediate consumption of the windfall."[184] The difficulty absorbing the windfall profits efficiently and the ensuing misallocation of production factors which in turn contributed to the failure to establish sustainable growth is known as 'the resource curse'. Also other mineral-producing countries suffer this 'curse'.[185]

The ultimate use a country makes of windfall gains and the benefits it derives from them depend on the ability of the country to formulate and execute a coherent strategy.[186] The formulation and execution of such a strategy depends "first, on the policy choices adopted by the producer government. These include not only extraction rates and public spending, but also possible shifts in trade and other policies in response to changes in national income, which affect the distribution of the gains across the economy. Second, they depend on how the macro-economy responds to the surge in demand, which will create disequilibrium in some markets at pre-windfall output and prices."[187] Other factors that influence the outcome of

[183] "(...) booming sector theory focuses on the sectoral reallocation of productive factors in response to a favourable shock emanating from either a resource discovery or an increase in the price of some commodity-typically an exportable one. If income is spent rather than saved abroad, the sum of the consequences includes a resource movement effect, which draws factors of production out of other activities and into the booming sector, and a spending effect, which draws factors of production out of activities producing traded commodities (to be substituted by imports) and into nontraded sector. The contraction or stagnation of the traded sectors is sometimes referred to as the 'Dutch Disease'." A. Gelb , *op. cit.*, p. 22.

[184] R.M. Auty, *Sustaining Development in Mineral Economies, The Resource Curse Thesis*, Routledge, London/New York, 1993, p. 17.

[185] R.M. Auty, *op. cit.*, p.8.

[186] A. Gelb, *op. cit.*, p. 37.

[187] *Ibid.*

the strategy are of a political and administrative nature: the stability of the government, which affects the scope of decision-making, and the level of centralization of decision-making, which determines the level of consistency of decision-making.[188] The potential contribution of oil to economic growth should be regarded within the institutional framework of an underdeveloped country that is constrained by bottlenecks arising from the shortage of certain key goods and services.[189] In response to the growing literature on industrialization and problems with structural change in oil-producing countries, Karshenas writes that: "It is important to distinguish between two aspects of the impact of government policy in such economies; namely the income-generating and the capacity-generating aspects."[190]

The Dutch Disease model is a static and single-period-only analysis. The construction boom theory is better suited to analyze the long-term effects of an income shock on savings and investments.[191] The effects of fluctuations in oil revenues in the years following the first oil windfall, the limited absorption capacity, and the effect on other sectors of the economy depend on the flexibility shown by governments in adjusting expenditure and distributing these effects across the economy. One important way of managing the economy is sterilizing the oil windfall. According to Auty, "A prudent government policy for the deployment of oil windfalls is premised on a clear distinction between the volatile and finite minerals sector and the sustainable non-oil economy. In order that the non-oil economy should escape severe distortion by over rapid absorption and/or withdrawal of the oil windfall it is necessary first to sterilize the windfall by allocating it to special development accounts. In the absence of viable domestic investment opportunities a significant fraction of the windfall may be held overseas to accumulate interest until barriers to effective

[188] *Ibid.*

[189] M. Karshenas, *Oil, State and Industrialization in Iran*, Cambridge University Press, Cambridge, 1990, p. 5.

[190] *Ibid.*

[191] A. Al-Yousuf, *Kuwait and Saudi Arabia: From Prosperity to Retrenchment*, O.I.E.S. Papers on Oil and Finance 7, Oxford Institute for Energy Studies, Oxford, 1990, p. 60.

investment such as transport bottlenecks and shortages of skills are removed."[192]

Bad management of windfall gains is partially responsible for the negative relationship between the abundance of natural resources and long-term economic growth, and for the disappointing performance of oil-producing countries.[193] Oil production is a high-rent activity, which fosters rent-seeking behaviour and corruption. However, the rent is not only a reward for controlling production, but also income received through the exploitation of social, political and legal privilege.[194] The importance of oil for the economy is therefore also determined by the social and political institutions it fosters.

The key problem in most oil-producing countries is that only a small elite, closely related to the government, is involved in the distribution of the oil windfalls. Although such a small group may initially be successful in getting investment projects off the ground, they are only able to manage a small number of very large (capital-intensive) projects. The oil boom of 1973-74 changed the prevailing notion of property rights, altered the relative power of interest groups and organizations, and changed the relationship of the state to the market.[195] The lack of governance (both corporate and political) will quickly lead to non-competitive behaviour and economic and political nepotism. The fact that the oil windfalls were combined with a move towards greater national or state control over the economy tended to limit international competition, throttle domestic competition and feed corruption. Leite and Weidman[196] show that the extent of corruption depends on the abundance of natural resources, government policies and the concentration of bureaucratic power. Intentionally or not, the institutional make-up of oil-producing countries after the first oil windfall was consistent with an economic and political system in which the state and certain elites frustrated the long-term efficiency of resource allocation.

[192] R.M. Auty, *op. cit.*, 1990, p. 93.

[193] C. Leite and J. Weidmann, *Does Mother Nature Corrupt? Natural Resources, Corruption, and Economic Growth*, IMF Working Paper, WP/99/85, p. 3.

[194] T.L. Karl, *op. cit.*, p. 6.

[195] *Ibid.*, p. 7.

[196] C. Leite and J. Weidmann, *op. cit*, pp. 30-1.

6.4 INVESTMENTS

In the low-absorbing economies of the Gulf states, the lack of infrastructure initially allowed the government to invest a large part of the windfalls in physical and human capital, without immediately having to pick 'industrial winners' that would produce sustained growth. Even at the best of times, it may be difficult for governments to select winners, due to informational asymmetries. However, since demand for infrastructure depended on other government investments rather than on demand from an existing private sector, the total planning process became inextricably intertwined. It was the bureaucratic planners at the ministries who were in command of the economy, not the market. This led to massive misallocation of factors of production, partly because the bureaucracy failed to formulate a coherent long-term strategy that was flexible enough to accommodate timely adjustments to fluctuations in the market, and partly because bureaucracy was not restrained by the political leadership from succumbing to the temptation to expand the public sector for reasons of patronage.[197] As a result, the return on investment was rather low.

The fact that the surplus was concentrated in the hands of the government and the planning of projects was rather knowledge- and labour-intensive typically led to large-scale capital- and energy-intensive projects rather than a large number of smaller projects.[198] The larger the project, the greater is the risk of large cost-overruns and delays in delivery. In Iran and Saudi Arabia, projects averaging over $1 billion were numerous.[199] Of the $1 billion-plus projects in countries such as Iran, Algeria, Venezuela, Mexico, Nigeria and Indonesia, 47% experienced cost escalations, completion delays and/or postponements and suspensions.[200] In the Gulf states, large-scale national industrialization projects resulted in regional overcapacity in shipbuilding, ship repairs, aluminium smelting, international airport capacity, and so on, until their governments decided to initiate the

[197] R. M. Auty, *op. cit.*, 1990, p. 96.

[198] J. Boers, G. Eisenloeffel, C. van der Linde, 'Olieraffinage en Petrochemie in de Golfstaten en EG', *ESB*, 22 February 1984.

[199] A. Gelb, The Oil Syndrome: Adjustment to Oil Windfall Gains in Oil-Exporting countries. In D. Lal and M. Wolf (eds.), *Stagflation, Savings and the State*, Oxford University Press, Oxford, 1986, p. 121.

[200] *Ibid.*, p.123.

regional coordination of projects by setting up the Gulf Cooperation Council (GCC). Both Kuwait and Saudi Arabia invested extensively in the refining and petrochemical industries. However, in these countries, too, the expected rents on cheap energy inputs were not enough to offset the disadvantageously high costs of plant construction, their remoteness from main consumer markets, and the trade barriers that their intermediate and final oil products encountered in consuming countries.[201]

In the mid-1980s, in all oil-producing countries, these industries were further disadvantaged by the vicious circle of falling prices, shrinking capacity utilization, and rising overheads. The fact that the new industries came on stream at a time when their main export markets were declining gave them little chance of overcoming the infant-industry stage and establishing themselves as competitive firms in the international markets. The contraction of economic activity also tied governments closer to their state-run industries. Since credibility of their economic and industrial strategies was at stake, governments were forced to continue to support these industries during the economic downturn, although the alternative income streams they had expected to receive failed to materialize.

The large investment commitments absorbed a substantial share of the oil windfalls. In Iran and Algeria, project costs were over 1% of 1977 and 1980 GDP, respectively. In addition, governments in these countries engaged in direct transfers and subsidies (explicit or implicit) to domestic consumers and producers, and public-works-facilitated employment programmes.[202] Dependence on oil revenues rapidly spread throughout the economy and reduced governments' ability to manage the economy later in the 1980s when oil prices fell. Oil revenues fluctuated, but much government expenditures had been committed for long periods of time. The fiscal health of most oil-producing countries deteriorated in the 1980s and eventually led to a sharp readjustment in later years. In the light of the cost- and time-overruns of the investment projects, together with the other problems of allocation and economic management, the decision to spend the oil windfall rather than save it, was, with the benefit of hindsight, not the most efficient solution.[203]

[201] J. Boers, G. Eisenloeffel, C. van der Linde, *op. cit.*, *ESB*, 22 February 1984.

[202] A. Gelb, *op. cit.*, 1986, Table 5-4, p. 123.

[203] *Ibid.*, p. 129.

To be sure, although certain countries, such as Kuwait and Saudi Arabia adopted different economic strategies, they agreed in deciding to accumulate foreign assets, particularly as long as the economy was unable to absorb the windfalls. Kuwait's rentier strategy entailed more prolonged dependence on imported of goods and skills. Saudi Arabia, on the other hand, chose to develop its productive capacity. Concerning the accomplishment of Saudi Arabia's strategy, Auty maintains: "In broadbush terms the strategy displays strong positive aspects. Absorptive capacity was expanded with speed and technical (if not economic) efficiency. Meanwhile the accumulation of overseas assets and liberal recourse to imported skills provided insulation against external shocks and permitted incremental adjustment when oil prices dipped in the mid-1980s. More detailed examination reveals less desirable features, notably a reluctance to broaden the tax base, increasing recourse to subsidies, and overexpansion of the bureaucracy."[204]

The problems of the high-absorbing oil-producing countries, who had financed a large part of their oil-based industrialization with foreign loans, were exacerbated by the costs of capital service charges.[205] The lack of competitiveness of the resource-based industries resulting from huge cost overruns was so serious that: "In the case of RBI (which was intended to provide an additional stream of export earnings), such cost overruns along with chronic overmanning and the failure of the expected resource rents to match expectations turned the projects into resource sinks which lost money rather than generated new foreign exchange, taxes and capital."[206] This resulted in greater dependence on income from oil rather than less. The only exception, arguably, is Indonesia, which due to coherent policies and an abundant supply of labour and other natural resources, as well as to certain institutional features of the economy, managed to escape the most serious aspects of the problems associated with the Dutch Disease.[207]

The oil-producing countries also suffered in that, like other countries, they were not isolated from the world economy. On the contrary, the collective action of the OPEC member states on the international oil

[204] R.M. Auty, *op. cit.*, 1990, p. 100.

[205] *Ibid.*, p. 252.

[206] R.M. Auty, *op. cit.*, 1993, p. 20.

[207] A. Gelb, *op. cit.*, (1988), pp. 223-4.

market threw into sharp relief the changing economic and monetary relations in the world. The recycling of petro-dollars tied the fate of the oil-producing countries to the developing oil-consumer countries, while the dollar exchange rate determined the value of both their income stream from oil and of their financial assets. Moreover, the wide fluctuations in the dollar exchange rate and interest rates in the 1970s and 1980s transmitted the wrong signals with regard to extraction rates, government spending, and any possible comparative advantages in capital- and energy-intensive industries. The windfall gains in 1973-1974 and again in 1979-1980, were also major external shocks to the economies of the oil-producing countries. The subsequent disequilibrium was misinterpreted as an alignment of the competitive advantages in the world, which only served to amplify the external shock of the collapse in oil prices and the resultant large loss of income in 1986. As the oil consuming developing countries had found earlier, it was very difficult to interpret the international economic indicators correctly and to know how the developing countries will respond, since the signals turned out to be full of 'white noise'. As it turned out, meeting the criteria with regard to management and strategy was not enough to offer protection against international economic developments or the developments on the international oil market. In sum, as Gelb puts it: "It is possible to conclude (...) that a great proportion of the potential gains to the exporters in this study [Algeria, Ecuador, Indonesia, Nigeria, Trinidad and Tobago and Venezuela] was nullified by a combination of the changes induced in the global economy by the oil shocks and the poor economic policies of the exporters themselves during the period. Together, these seriously reduced the efficiency with which the countries used their resources, as judged by their own criteria of diversification and growth."[208]

6.5 Oil and Development

Oil income has fluctuated widely since the 1970s, and this, with hindsight, contributed to the inefficient way in which the investment plans were implemented. The value of OPEC oil exports increased as a result of two oil price rises from a low $23 billion in 1972 to $284 billion in 1980. As it turned out, 1980 was a peak year for the OPEC member states in terms of the value of oil exports. Due to a drop in oil prices in the mid-

[208] A. Gelb, *op. cit.*, 1988, p. 143.

1980s, the value of exports fell to a low $77 billion in 1986, rising gradually to $128 billion in 1991.[209] In 1998, when oil prices fell to an average price of about $12 a barrel, OPEC oil export revenues were around $120 billion.[210] Fluctuations in the value of oil exports of individual member states depended to a large extent on their share in OPEC production agreements. Although some countries, like Saudi Arabia, imposed more stringent restrictions on production in the 1980s in order to maintain the preferred OPEC price, all countries suffered from very wide fluctuation in income. Saudi Arabia's government oil revenue has fluctuated widely, from a high $97 billion in 1982 to a low $19.5 billion in 1988 and $36.3 billion in 1998. As a result, the Saudi government's budget has not been in surplus since 1983.[211] In 1987, the country's fiscal deficit had increased to 25.3% of GDP, since which time the government has managed to bring it down 2.9% (1997).[212] The price collapse in 1998-1999 threatened to thwart the efforts to bring down the fiscal deficit, because 1998 closed with an expected deficit of 9,5% of GDP.[213] The importance of oil to the Saudi economy is further illustrated by the government's budget for oil revenues. The contribution of oil to government income has decreased from 98% in 1982 to approximately 78% in 1998.[214] The foreign currency receipts are

[209] *OPEC Annual Statistical Bulletin*, 1991, Table 5, p.6. The OPEC total does not include Kuwait and Iraq. In 1990, the total value of OPEC oil exports was $146 billion.

[210] No official data are available yet. This figure has been calculated as follows: export volume estimate based on 1998 OPEC production of 30,730 thousand b/d (*BP Amoco Statistical Review* 1999, p. 7) minus domestic consumption times average price level of $12 per barrel. In March 1999, OPEC was reported as having produced 27,640 b/d (*Barron's*, May 17, 1999). Oil export revenue in the first quarter of 1999 was therefore substantially lower than in the same period in 1998. Since the oil price declined further after August 1998, and the 1998 production cutbacks did not work, oil export revenue continued to decline. The March 1999 effort to curb output is easily explained by these data.

[211] Saudi Arabian Monetary Agency, *Thirty-Fourth Annual Report*, 1419H (1998G), p. 21 and pp. 306-7.

[212] *Ibid.*, p. 125.

[213] M. Zonis, 'The New "Oil Shock", Political Instability is Next', *PIW*, February 1, 1999, p. 7.

[214] Saudi Arabian Monetary Agency, *op. cit.*, p. 124.

substantially dependent on oil, and the oil sector contributes 39% to GDP (at current prices).[215]

The dependency of the government on oil is also large in other oil-producing countries, despite efforts to diversify. In Algeria, government revenues depend for 58% on oil, while oil and gas accounted for 97% of total export revenues. In Indonesia, oil and gas export account for 23% of total export revenues. Iran's government revenue depend for about 36% on oil revenues, while oil and gas contribute about 80-85% to total export earnings. Lastly, in Nigeria oil and gas accounts for 60% of government revenues and 95% of the foreign exchange earnings.[216]

The achievements in economic growth have been disappointing. More than twenty-five years after the oil price increase of 1973-1974, in terms of GNP per capita, seven OPEC member states fall in the category of low-income countries (per capita income of $765 or less; Nigeria) or lower-middle-income countries (per capita income between $766 and $3,035; Indonesia, Iran, Iraq, Ecuador, Algeria and Venezuela). Another three member states rate as upper-middle-income countries (between $3,035 and $9,385; Gabon, Saudi Arabia and Libya); and only the United Arab Emirates, Kuwait, and Qatar count as high-income countries ($9,385 or more).[217]

6.6 CONCLUSION

The disappointing outcome of the attempts to absorb windfall profits is due to several factors. Firstly, like other developing countries, the oil-producing countries experienced a number of severe external shocks that disturbed their macro-economic stability. The combination of a drop in the price of oil and the depreciation of the dollar in 1985-1986 severely worsened the terms of trade of the oil-producing countries. Those oil-producing countries that had supplemented their oil income with international loans were subsequently hurt by the relatively high interest rates in the early 1980s. Secondly, industrialized countries continued to protect their markets. Negotiations between the EC and the countries of the GCC on reducing tariffs of petrochemical products, for instance, were dragged

[215] *Ibid.*, p. 301 and p. 274.

[216] M. Zonis, *op. cit. PIW*, pp. 6-7.

[217] *World Bank Development Report 1997*, Table 1, pp. 214-5 and pp. 264-5.

out until the European industry had been properly restructured.[218] Furthermore, the increasing use of non-tariff barriers and subsidization in the 1970s and 1980s created unfair competition in these important (and new) export markets for the oil-producing countries. The final factor that contributed to the disappointing results of absorbing the windfall gains was the inefficient nature of macro- and micro-economic government policies. The two oil price increases led to irrational oil-market expectations, subsequently thwarting the income expectations and development planning of the oil-producing countries. These problems were exacerbated by the continuing weak oil price in the 1990s.

Efforts to stabilize oil prices failed because OPEC, despite its large reserves, did not represent a large enough market share in world production to really make a difference. The existence of enough spare capacity, including that which was outside OPEC, thwarted every effort to limit production. Non-OPEC producers could always immediately increase production, and the production policy frequently failed because OPEC member states themselves could not live up to the production agreements. This continuous and widespread free-riding both by non-OPEC and by certain OPEC countries made any attempt to stabilize prices costly and ineffective effort. Moreover, Saudi Arabia, when its own oil income became insufficient to cover domestic government expenditures, was increasingly less able and willing to assume the cost of production reductions.

The bad management of windfall gains, on top of the effects of the Dutch Disease on the economy, contributed to the disappointing performance of the OPEC economies. The enormous rent-seeking opportunities were translated into economic and political nepotism and harmed the growth prospects of the economies. The aftermath of the Asian financial crisis, in which nepotism was unmasked in various countries as a serious restraint on economic growth and the distribution of wealth, has led to political changes in some countries. However, the outcome of the reform process in Indonesia and Venezuela, for instance, is by no means certain. In Indonesia, elections have weakened the government and the centralized corruption of the old regime may very well develop into decentralized

[218] J. Boers, G. Eisenloeffel, C. van der Linde, *op. cit.*, *ESB*, 22 February 1984.

corruption. In that case, the country will be worse off than ever.[219] In Venezuela, although Chavez was elected under a strong anti-corruption ticket, the change in government may merely constitute a change in rent-seeking elites rather than a real break with corruption. Fears among international investors concerning the changes in Venezuela began to grow as the meeting of the constituent assembly in July 1999 approached.[220] Chavez countered these fears by stating that he wanted to transform the political regime not the economy, and that he would introduce legislation that would guarantee foreign investors judicial security and the ability to repatriate capital.[221] However, the appointment of army officers to key economic posts in the government and in important state companies, such as PDVSA, is not reassuring and points to a change in elites rather than a serious attempt to deal with nepotism.

In order to break out of the vicious circle, countries will have to strengthen their institution building, regulate the economy rather than operate it, and truly open up to international competition. Liberalization and privatization, if properly introduced and monitored, may be able to break the cycle of corruption, and realize the oil-producing countries' potential for economic growth.

[219] See Chapter 3.

[220] *Financial Times*, 12 May 1999.

[221] *Financial Times*, 12 May 1999.

CHAPTER 7

NATIONAL OIL COMPANIES

7.1 NATIONALIZATION

The property rights associated with the large oil-rich concessions in the Middle East had allowed the large international oil companies to capture a substantial part of the economic rents. The appropriation of these rents was the priority of the governments in major oil-producing countries. In most OPEC countries was the favoured method to capture these rents nationalization (or at least partial nationalization). Given the dominant economic views in the early 1970s, the governments presumed that only nationalization of the concessions of the international oil companies would shift the entire economic rents to the state, which in turn would provide them with the unique opportunity to invest these revenues in the development of the economy.

The reason why the governments of large oil-producing countries believed that nationalization was the optimum strategy and that taxation was not an option was that, on the one hand, the government could profit politically from nationalization and, on the other hand, wanted access to necessary information. This information was locked inside the companies and not easily available or retrievable. The international oil companies were vertically integrated firms, operating in many countries and different parts of the value chain. The strategic core of an oil company, represented in assets of high specificity, is of great importance to the achievement of their strategic goals. The specific availability of natural or technological resources, human assets and investments, know-how and the firm's particular internal governance of these assets create the rationality for the firm's existence.[222] There are four types of asset specificity that create core skills: (1) site specificity originating in resource immobility; (2) physical asset specificity as a result of technological advantages; (3) human asset specificity as a consequence of know how advantages; and (4) dedicated

[222] Torger Reve, 'The Firm as a Nexus of Internal and External Contract' In *The Firm as a Nexus of Treaties*, Masahiko Aoki, Bo Gustafsson and Oliver E. Williamson (eds.), Sage Publications, London, 1990, p. 139.

assets as a result of specialized investments.[223] The fact that, before 1973, the international oil companies functioned in relative isolation from the rest of the domestic economy in oil-producing countries, that little local labour was employed (particularly in higher management positions) and that domestic firms had either only a small presence or none at all in the sector made it impossible for the national governments to gain access to the information they needed. Moreover, the international oil companies owned the property rights to the oil resources and, although the dedicated assets were relatively low compared to their investments in the consumer markets, they had technological and human asset advantages over the local producers and the government. The oil companies therefore did not want to share information with the government or local firms, since it would impede the achievement of their strategic goals. Furthermore, the information gave them a strong bargaining position in their dealings with the government.

Local knowledge of the oil industry in the oil-producing countries was fairly small and, for instance, increasing the participation of the local labour forced in the oil sector was not enough to acquire more and superior information. In this light, the creation of OPEC in 1960 must also be regarded as an attempt to gain more information from the oil companies and to obtain a better bargaining position in the governments' dealings with them. Nevertheless, the results of the governments' efforts to gain access to the oil industry in their countries remained limited. In 1973, despite the favourable oil market conditions for OPEC countries to assert a stronger bargaining position vis-à-vis the oil companies, they could only exert their increased power through the appropriation of oil assets.

At the time, the decision of the governments of the OPEC member states to bring, to a varying degree, the oil industry under state control, and thus to create a radical break with the earlier organization of the property rights in the oil sector was seen as the only possible way to gain information. The change of property rights dislodged the oil companies from their site-specific assets and released the companies from their earlier position in which they would not disclose information. The bargaining position of the governments had been completely transformed due to the transfer of property rights. The oil companies, knowing that the nationalization of

[223] P. Milgrom and J. Roberts, *Economics, Organization and Management*, Prentice Hall International, Edgewood Cliffs, 1992, pp. 106-8.

their assets was final, were now in a position to favour disclosure of information in order to gain a proper compensation for their nationalized assets.[224] The sheer size of the nationalized oil assets, and, in the short term, the lack of alternative sources of crude oil was a large incentive for the international oil companies to cooperate with the governments after nationalization or partial nationalization. This also allowed the governments of the oil-producing countries to negotiate the oil companies' assistance in transferring knowledge to the, in some cases newly created, national oil companies.[225]

The need for information on the operations of the international oil companies was not limited to the OPEC member states. After 1973, European governments also decided to increase direct state participation in the oil sector based on the argument that the regulations valid at the time, were insufficient to gain accurate information. However, governments can obtain any information they want by regulating the industry. According to Grayson[226], the problem of the European governments was not so much their inability to obtain information but their inability to properly analyze the information: "Thus the NOC's aided their govern-

[224] In 1974, Kuwait purchased 60% of KPC for $112 million, based on the net book value. The remaining 40% would, on the same basis, cost around $70-80 million. However, in 1975, the Kuwaiti Oil Minister Abd al-Muttalib al-Kazimi stated that the compensation for the remaining 40% would not be on the same scale as before. At the time, it was estimated that the compensation would only be around $35 million, half the amount of the former net book value. *Petroleum Economist*, April 1975, p. 124.

[225] In Venezuela, compensation for nationalizing the oil industry was initially established on $1.04 billion. The companies initially received $475 million, while the remainder was placed into a guarantee fund and would be paid when installations had been inspected. The government claimed that the installations were not in good working order and deducted $135 million from the fund. In 1976, the remainder in the fund ($425 million) was frozen when the government sought $635 million from companies in unpaid taxes. The companies offered to settle for $135 million, but the government refused to accept this offer and the parties went to court. L. Randall, *The Political Economy of Venezuelan Oil*, Praeger, New York, 1987, pp. 55-6; As one of the sources of PDVSA's capital is listed: Bs 6,165 million for the capitalized (book) value of fixed assets and indemnified materials from the concessionaires. The book value was less of one-fifth than the market value. *Ibid.*, p .46 and note 15, p. 60.

[226] L.E. Grayson, *National Oil Companies*, John Wiley and Sons, Chichester, 1981, p. 10.

ments' understanding of the oil industry by interpreting oil industry information -whether that information was provided by themselves or by the MNC's. Governments hoped that having a window on the oil industry would enable them to establish 'a yardstick' against which they could judge the performance of the oil MNC's in their country." Grayson goes on to argue that in order to make the 'yardstick' work, the NOCs must work in the same way as the large international oil companies. In other words, the NOCs should not receive government assistance, nor should they be forced to perform non-commercial activities. In most cases in the oil-producing countries, the NOCs did indeed receive assistance and were also forced to perform non-commercial activities, which made their yardstick function worthless. Moreover, the NOC should also properly inform the government and not act as an agency with its own rent-seeking motives. The incentives in state firms to reduce costs are not very high, and, therefore, their performance cannot be measured against international oil companies.

The information issue was very important to OPEC member states. Particularly in the countries around the Persian/Arabian Gulf, the international oil companies enjoyed a very strong position in the oil sectors in these countries. The sheer size of the concessions made the barriers for entry very high and virtually impossible to overcome. The local private capital markets would have been too small to participate in the domestic oil industry had that opportunity been available. The price increases of 1973-1974 elevated the importance of the oil industry to the countries' economy. Export income was very much dependent on earnings from oil exports, and the contribution of oil to the governments' budgets was very large. As a result, the incentive for governments to increase control over the most important sector of their economy was very strong. Since in most OPEC countries foreign capital could not be replaced with local capital and given the historic economic and political exploitation, the government decided to take a leading role in this sector of the economy.

The government had another justification for nationalization in the potential disclosure of information about the oil industry. Organizing the oil resources in national oil companies created a situation in which local knowledge could be increased by learning-by-doing. In countries with a small labour market, such as Saudi Arabia, Kuwait, the UAE and Qatar, the learning process was necessarily limited to management and engineering skills, while for operational jobs, labour was predominantly imported from other Arab and Asian countries. In the other oil-producing countries, the learning process at operational level was already more diffused than in the Gulf countries, because more local labour had been employed by the international companies. At the level of managerial skills, these countries also lacked proper skills. Except for Venezuela, where a process of Venezuelanization in the 1960s had provided for a substantial the local content of management.[227] In general, the development of a broad technological and human capital base requires large investments in education, labour market diversity, and organizational structures that will not only allow the production and implementation of industrial designs, but also quick diffusion of knowledge in the economy. The implementation of these information structures takes time and can only succeed with the proper incentives in place.

Although the appropriation of property rights by the OPEC member states gave them more information than they had before, the disclosure of information could not be complete. Through their international and vertical integrated organization structure, the international oil companies possessed firm-specific information that could not easily be transferred to the oil-producing countries. This information was embedded in the human capital and technology of the firms, but it also concerned information that resulted from the specific organizational and capital structure of the firms. In the 1970s, the core skills of the national oil company were limited to site-specific resources and some dedicated assets, while the technology and human capital core skills were still mainly underdeveloped. The pressure to develop these skills was large, since the companies had to operate in markets where other companies possessed these skills and the accompanying information rents.

However, nationalization did not completely solve the information problem of the government. The incompleteness of information can also be reinforced by the NOC itself, when it does not disclose information because of rent-seeking behaviour. Venezuela's new president, Chavez, has repeatedly accused PDVSA of being a state within a state, and blamed executives of lavish spending.[228] In an effort to gain control over the company, Chavez has appointed some of his own people to run PDVSA's

[227] L. Randall, *op. cit.*, pp. 92-3.

[228] *Financial Times*, 10 March 1999.

finance, personnel and planning departments. This has increased the fears of more government intervention in the oil sector.

7.2 THE AUTONOMY OF NOCs

In a country endowed with substantial oil resources, the state has several interests to consider and protect. These interests range from security of supply on the domestic market, health, safety, welfare and environmental interests, the conservation of oil resources, the training and employment of local labour in the oil industry, and finally, the generation of proper returns on the exploitation of oil resources for the state.[229]

If government, in the 1970s, had been solely interested in modifying the profit-making goals for reasons of national interests, and if governments had been able to obtain proper information, regulatory reforms of the industry would have sufficed. Moreover, the governments recognized that the international oil companies gave them access to markets, technology and skills.[230] Each state therefore designed its petroleum regime to suit its economic and political needs. If the state wants to continue to attract foreign private capital in the oil sector, it should design a regime that provides enough security to foreign companies and bankers to warrant large and risky investments in their petroleum sector. Under such a regime, the state accepts that it will share some of the rents with these investors. If a country wants to reduce (as compared to an earlier period) and/or control the exploitation of their oil resources, it might prefer a less attractive regime for private foreign investors or limit investments in the sector by other measures. This is particularly the case when the state expects the appropriation of the economic rents to measure positively against the entrepreneurial risks and the price of capital and technology. In the 1970s, the larger oil-producing countries opted predominantly for full state ownership of the oil industry.

The drive of governments to organize their oil sector in national oil companies was inspired by several factors:

- the shared bad experience by developing countries with multinational companies;

[229] R. Bentham, 'Legal Status of State Petroleum Companies' In N. Beredjick and T. Wälde (eds.), *Petroleum Investment Policies in Developing Countries*, Graham & Trotman, London, 1988, p. 258.

[230] L. Randall, *op. cit.*, p. 41.

- too large commercial risks for the (small) domestic private sectors to bear;
- the fact that shared state-domestic ownership would have made taxation of the oil industry harder and would have reduced the governments' income[231];
- the ability to contain possible divergence between state interests and private corporate interests[232];
- the ability to implement a more effective energy policy and environmental protection[233];
- the larger benefits of the learning process and/or the broadening of corporate objectives to socio-political goals;
- the facilitation of technology transfer within the economy;
- the possibility to extend their political patronage[234];
- the fact that profits from the nationalized oil industry "could be disposed of directly by public authorities in the public interest"[235] and could help to establish linkages with the other sectors of the economy[236];
- the fact that control of the country's major source of foreign exchange earnings would be a safeguard for securing foreign credits (a major function of oil income)[237]; and
- to ability to acquire knowledge from the oil industry.[238]

The importance individual governments attached to each of the above factors determined the relationship between the government in question and the national oil company. The OPEC NOCs were either created by public act, statute or decree as a public corporation or a joint stock corporation owned by the state. The level of state intervention and the legal structure of the NOC depended on the dominant political and economic regime, the

[231] L. Randall, *op. cit.*, p. 40.

[232] R. M. Auty, *op. cit.*, 1990, p. 134.

[233] J.M. Griffin and H.B. Steele, *Energy Economics and Policy*, Academic Press, New York, 1980, p. 283.

[234] R. M. Auty, *op. cit.*,1990, pp. 134-5.

[235] J. M. Griffin and H.B. Steele, *op. cit.*, p. 283.

[236] D. Fee, *op. cit.*, p. 59.

[237] M. Karshenas, *op. cit.*, p. 171.

[238] D. Fee, *op. cit.*, p. 59.

importance of the oil sector to the domestic economy and the government's budget, and the availability of management and labour. The NOC was never intended to be just any oil (and gas) company with a narrowly defined core business. Fee rightly argues that: "Whichever legal structure is adopted, recognition must be given to the dual nature of the NOC as an entity conducting its activities like a commercial enterprise while at the same time carrying out services demanded by the state."[239] These services to the state included "redistributing income, subsidizing particular regions and sectors, earning foreign exchange, generating employment, and increasing the probability that the party in power will be re-elected."[240] The state had given the companies the task to operate their rights over petroleum and in certain cases also the downstream industries, and, in addition, to carry out other objectives of the state. The NOCs were the governments agents. The degree of direct control of the governments varies, but all OPEC governments exert some influence through the appointment of one or several members of the Board.

In the case of Venezuela, the overall direction and administration of the company is in the hands of the Assembly, which is presided by the Minister of Mines and Hydrocarbons. In the UAE, Saudi Arabia and Kuwait, the Minister of Energy or Petroleum and/or Mining is Chairman of the Board. In Ecuador, Nigeria, Libya and Iraq, the important ministries appoint or propose a representative to the Board of Directors. In Indonesia, the Chairman of the Board is appointed by the President, while in Qatar, all members of the board are appointed by the amir.[241]

Certain investment projects (e.g., prospecting and exploration in Abu Dhabi), joint ventures and foreign loans are subject to government approval, and the budgets of the companies also need to be approved by the government. Although the organizational structure of the companies plays an important part in demarcating the boundaries between government and company in law, the political structure, the economic welfare of a country and/or the type of government has, in practice, been a decisive factor for

[239] D. Fee, *op. cit.*, p. 62.

[240] M.J. Horn, *The Political Economy of Public Administration*, Cambridge University Press, 1995, p. 136.

[241] R.W. Bentham and W.G.R. Smith, *State Petroleum Corporations: Corporate Forms, Powers and Control*, Centre for Petroleum and Mineral Law Studies, University of Dundee, Scotland, 1986.

the level of autonomy of the companies. In Venezuela, PDVSA used to operate as a fairly independent company until the government ran into serious debt-servicing trouble and disagreed with the company's expansion plans that required substantial foreign borrowing. In Kuwait, the government is closely involved in the formulation and execution of KPC's strategies, both national and international, but major disputes between the government and KPC have not surfaced. In Saudi Arabia, where four state companies were involved in the oil industry (Petromin, Sabic, Samarec and Saudi Aramco), there was a struggle in the early 1990s between these companies (and their management and their supporters in the government administration) for supremacy, in which Saudi Aramco prevailed.[242] "The Gulf War also served to strengthen Saudi Aramco in its bid to maintain its primacy in the Saudi oil industry over the recently-formed marketing and refining subsidiary Samarec.(...) Local industry officials say that while Samarec managed to keep the allied war effort supplied with fuel, enough cracks appeared in its system to sow doubts about the overall quality of its management performance. Nazer has always regarded Aramco as his personal baby, so his strength and Samarec's weaknesses have only helped Aramco's position, in particular with regards to downstream investments."[243]

For the NOCs, the level of autonomy granted to them by the government determines their ability to implement appropriate market strategies and to apply efficient management methods. In the absence of proper independence for the firm, the government can force it to serve the government's employment, welfare, trade, industrial, monetary and energy policies. For instance, the government could require the firm to use more local technology, equipment or suppliers, or it could hamper the companies' desire to enhance productivity by reducing staff.[244] The government can also impose its domestic economic strategy on the NOC, which may increase production costs and reduce the international competitiveness of the firm.

NOCs may be subject to limitations with regard to which part of the value chain they are allowed to engaged in (e.g., oil exploration and crude

[242] King of Oil Surges Ahead, *Petroleum Economist,* December 1991, p. 6.

[243] *Ibid.*

[244] J-P Anastassopoulos et al., *op. cit.*, p. 38.

production, refining, marketing and/or petrochemicals). The state can also limit the company's geographical scope (for instance, to the domestic market or a part thereof) and/or the company can be restricted in the types of legal agreements it can enter into with companies (i.e., private companies or foreign companies). Finally, the state can impose restrictions on how or to what extent the company raises capital in private capital markets.

NOCs usually have monopoly or oligopoly positions in their domestic markets with regard to the production, processing and distribution of oil and oil products. This structure of the domestic market creates barriers to entry. In crude oil exploration and production, these barriers are enormous. In refining, the barriers to entry are also large, although some countries, in particular Saudi Arabia, have established joint ventures in refining. In petrochemicals, the barriers to entry are less high, but as in the other part of the value chain, where private participation is allowed, the NOC or other state company insists on holding a majority share.

The restrictions on the companies' geographical scope is sometimes an issue of major debate between the NOCs and the governments. Some NOCs, notably KPC and PDVSA and also Saudi Aramco, have initiated international investment strategies as a result of which they no longer concentrate their investment efforts solely on the domestic economy. From a company point of view, securing markets for crude oil in jointly-owned refineries in oil-consuming markets can be an efficient strategy. However, the government can take the position that, even though the company has used capital raised on foreign markets, the outstanding liabilities could limit investment of the company on the domestic market. International strategies also limit the significance of the local economy to the company. Although this is positive for the firm, it can be a disadvantage to the government and it diminishes the governments's control over the NOC's operations. The limitations on the type of legal agreements they are allowed to engage in can seriously hamper the ability of the NOC to get access to capital and technology.

The financial structure of national oil companies can also limit the functioning of the firm. In general, every firm has certain resources available for expanding the finances of the firm. The firm can seek additional capital from the shareholders, issue bonds or attempt to obtain long, medium or short-term credits from banks. Since typically the only shareholder of the NOC is the state, the state can raise the firm's capital (for which provisions have been made in the statutes of most of the OPEC

NOCs[245]) or by securing a certain budgetary endowments. However, securing such additional capital through a budgetary grant is not so easy. The demand from other state companies and government departments is large, and particularly when the government is running a budget deficit, an additional call on the budget is unlikely to be rewarded. Borrowing from private investors through bond issues or bank credits usually needs to be approved by the government, although some NOCs, like KPC, have been given discretion in these matters. If the government does not wish to increase the NOC's capital and the firm resorts to borrowing, the financial structure of the firm may become unbalanced. However, the biggest constraint on acquiring capital on the international capital markets, is the foreign debt of the country. In December 1998, the notes of PDV Finance (an offshore investment vehicle of PDVSA) were downgraded to A-, while the long term currency debt of the country was downgraded from BB- to B+ because "the structure of the PDVSA Finance transaction mitigated but did not completely eliminate sovereign risk."[246]

The scarcity of foreign exchange has stimulated treasuries in oil-producing countries to demand the surrender of foreign receipts. For instance, the decree of 25 September 1982 in Venezuela required all state enterprises to give their foreign exchange to the Central Bank.[247] When the price of foreign exchange in Bolivars increased, and domestic interest rates did not increase enough to offset the loss, this meant that the deferred interest earned on foreign exchange was smaller and the anticipated income was reduced.[248] Such foreign exchange restraints can also limit the NOCs in making the necessary foreign purchases, because the Central Bank or the government needs to preserve hard currency reserves for other sectors of the economy or for debt servicing. As a result, in countries with serious debt problems, the NOC may not be able to obtain the appropriate capital goods or spare parts and could be forced to use domestically produced goods instead. This is no problem if the price of these domestic goods properly reflects quality and if the goods are available. However, if these domestic goods do not match the foreign goods (in price, quality or tech-

[245] R.W. Bentham and W.G.R Smith, *op. cit.*, pp. 10-97.

[246] *Financial Times*, 10 March 1999.

[247] L. Randall, *op. cit.*, p. 47.

[248] *Ibid.*

nology), the NOC will be forced to find another (less preferable) solution and/or run down certain production facilities. The Russian oil industry in the late 1980s and early 1990s presents an extreme example of such a situation. The ethnic conflict in Azerbaijan in the late 1980s crippled the supply of oilfield equipment. Azerbaijan supplied around 60% of the domestic needs. The replacement of domestic goods by imported goods was limited by the scarcity of foreign currency, and the Russian oil industry could not maintain production levels without new equipment. In this case, the oil engineers performed a miracle in that they managed to keep production going for as long as they did. However, the collapse, when it came in the early 1990s was all the more severe.[249] Maintenance of the production facilities has now also degenerated in other oil-producing countries, thus intensifying the downward spiral of the domestic oil industry.

In the 1990s, NOC executives across the oil-producing countries have repeatedly complained that too little revenue is ploughed back into the NOCs to maintain the national oil industry, let alone expand it. The NOCs' revenues are determined in part by the production quota and oil price levels their governments manage to secure or sustain through OPEC. Such an agreement depends not only on commercial grounds, but also on the political position of that member state in the organization. The March 1999 production agreement again illustrates the political nature of the distribution of production cuts. Allowances have to be made for the demands by certain member states (e.g., Iran), and the difference has to be made up by the other member states.[250] The NOCs lack the powers of decision-making that their counterparts in a private international oil company can exercise with respect to industrial and commercial operations.[251] In a competitive market, this will seriously hamper management to act.

Foreign direct investments by the NOCs in a country with an open and abundant capital market can help to obtain additional investment capital, since part of the capital requirement can be borrowed on the local markets. Joint ventures are also a vehicle to raise capital on foreign markets. Not only does the partner firm contribute to the capital, it also allows the NOC

[249] C. van der Linde, Empty Barrels Make the Most Sound, *World Competition*, Vol. 15, no. 2, December 1991, pp. 152-3.

[250] J. Richard, *op. cit.*, p. 43.

[251] J.E. Hartshorn, *op. cit.*, pp. 165-6.

to borrow on the local market and/or the joint venture can be listed on the local stock exchange. However, governments usually oppose to too much diluting of the capital of their subsidiaries, because it reduces government control over the firm.[252] In general, governments of developing countries with economies that still require substantial capital formation and with budget deficits limiting their possibility to invest, view with disapproval the investments made abroad by the (more successful) state oil companies. They view these investments as investments withheld from the domestic economy. Although investments in a foreign country and/or in the downstream industry give the NOC a better return on capital than investments in the domestic economy, the government usually judges these firm strategies as running counter to economic policy.

Again, the discussion of the expansion plans of PDVSA in Venezuela and abroad is an excellent example of how interests differ for the NOC and the government. First, tensions between the government and the company arose over the foreign expansion plans in consumer markets. Later, tensions arose when PDVSA favoured foreign direct investments in oil exploration in Venezuela. The expansion into new oil deposits required foreign capital, and PDVSA wanted to be able to develop these fields in joint ventures. Since the election of Chavez in 1998, the expansion plans of PDVSA have been under pressure. The Chavez government has now imposed an austerity package on the company, entailing a cut in expenditure of more than 36 percent of its 1999 budget.[253] Reportedly, the government not only imposed these measures to cut costs (which was necessary because of the low oil prices), but also to try to downscale PDVSA to free more resources for non-oil economic activity. Although the imposition of allegedly hard budget constraints on the company might induce greater efficiency and a reduction of the substantial overheads, the effect of such budget constraints will depend on the government's commitment of the government to restrict the company's pursuit of inefficient activities.[254] The fact that Chavez proposed to include Cuba in the San Jose Pact (under which 11 Central American and Caribbean countries buy oil at normal

[252] J-P Anastassopoulos et al., *op. cit.*, p. 71.

[253] *Financial Times*, 10 March 1999.

[254] Y. Qian and C. Xu, 'Innovation and Bureaucracy Under Hard and Soft Budget Constraints', *Review of Economic Studies*, 1998, pp. 156-9.

market prices but can borrow up to 20% of the value of the contract at discounted interest rates), does not bode well.[255] Fortunately for PDVSA, Mexico vetoed that proposal. Cuba is $68 million behind in repaying its oil debt to Mexico and $60 million behind in repaying its debt to PDVSA. A PDVSA official was quoted saying: "We have little to say about the policy. The president makes the policy. But we would be opposed".[256] This recent interference in the company's business is further evidence that the relationship between the government and PDVSA is not particularly cosy.

Chavez' attempt to reduce rent-seeking by company executives seems to be of questionable sincerity and could be intended merely to change the group of beneficiaries or the means of rent-seeking in the company. His political rhetoric prevents him, for instance, from extending the policy of privatization to include PDVSA in the near future, since it is probable that predominantly foreign investors would benefit from any such a flotation of the company. The 1998 corporate restructuring plan to make 14,000 of the 42,000 workers redundant within the next ten years is already triggering political opposition from labour unions.[257] The government's commitment to its austerity measures might accelerate these redundancies, and political opposition and labour unrest might grow. The government's close relationship with the military might tempt Chavez to take full control and break off the liberalization process.

In general, the NOC's lack of autonomy over its corporate finances may create long-term problems for its very existence. This problem, which most NOCs have to face, jeopardizes its future as a firm capable of competing on the international oil markets. Concerning PDVSA's financial situation, Randall observes that "Although PDVSA did not have autonomy in the management of its finances, it was and is expected to continue to be self-financing. Self-financing might not be possible in the long run because the government has levied heavy taxes on PDVSA's income and has required PDVSA to price its products on the domestic market at levels below those on the world market and at times below cost, further reducing PDVSA's income below levels it would have obtained in a free market. Thus, although centralization of financial management brought the affiliates

[255] *Miami Herald*, 6 August 1999.

[256] *Miami Herald*, 6 August 1999.

[257] *Financial Times*, 10 March 1999.

under PDVSA's control, it did not provide PDVSA with the financial independence sometimes attributed to state-owned enterprises in general but rarely permitted to large state-owned oil companies."[258]

The attempts of PDVSA's management over the past ten years to develop a company strategy more independent of the government cannot be seen as an attempt to create a more efficient and competitive company only. Indeed, in the minds of PDVSA executives, that was all they were doing: "This was particularly true of Petroleos de Venezuela where we have spent the last 15 years reshaping our industry and our corporate structure from that of a State oil company into a global energy corporation. This is because we understood that in order to survive and to compete in the changing world we had to move away from the *restrictions of nationalism* [emphasis supplied], to take full advantage of the growing potential for internationalism."[259] Chavez, however, seems to disagree with this appraisal of PDVSA's corporate strategy. The mildly harder budget constraint imposed by Chavez would only serve to underpin that effort and yet they are greatly opposed by management. However, if these hard budget constraints are imposed without allowing the company to disengage from non-profit making activities and services to the state (that is to say commercializing it), the austerity measures will only result in the run-down of capital goods. The latter seems to be the concern of the former PDVSA president Pietri. In response to the austerity measures, he said, "It is being run down. The government is decapitalizing the company. If things continue at this rate, the company is going lose its competitiveness."[260] PDVSA's earlier brief period of independence from the state must also be seen as an attempt to appropriate more of the rents for the company and its management. The struggle for these rents has reached a new level since the Chavez government took office in December 1998. The struggle for power over the company reached a highpoint when the president of PDVSA was replaced in August 1999.

[258] L. Randall, *op. cit.*, p. 48.

[259] A. Casas Gonzalez (Man. Director PDVSA UK), 'How National Oil Companies are Meeting the New Conditions', Montreux Energy Roundtable 1991, New Management and Investment Challenges, 17-19 April 1991, p. 1.

[260] *Financial Times*, 10 March 1999.

7.3 Taxation

As was pointed out in the previous section, the lack of autonomy of the NOCs, particularly with regard to corporate finance, can have a large impact on the future of the firm. The oil regimes implemented in the 1970s were perceived as appropriate for the governments' goals at that time. However, the increasing price volatility of oil has rendered certain policy instruments inappropriate. The failure to efficiently absorb the oil windfalls, the subsequent economic instability during the downturn of the cycle, and increasing uncertainty may all have an effect on how private (international) oil companies assess the risks. If they perceive the risks to be higher this may induce them to increase the risk premium in their required expected returns. Because oil-producing countries relied heavily on taxation and quasi-fiscal instruments (such as state participation) for their rent collection, they denied themselves a flexible response to the changing investment environment and consequently hampered the infusion of skills and capital.[261] Conventional fiscal instruments, such as royalties, income taxes and PSCs, are sub-optimal tools for collecting economic rents in that they are too rigid to be applied in a dynamic international oil market. Resource rent taxes, whereby the investor is allowed to recover his investment and earn a threshold rate of return before tax is applied, do not suffer from this inflexibility.[262] However, this tax still implies substantial risk-sharing for the government in situations of oil-price volatility and cost variation among fields, although it does provide a stable contractual environment for oil exploitation. Despite its long-term advantages, such a tax has not been widely introduced.

Most OPEC countries where international oil companies were active (e.g., in Nigeria, Ecuador, Indonesia, Iran, Algeria and for heavy oil only Venezuela) modified their oil-tax regimes but they generally refrained from implementing a resource rent tax. The oil regimes in the Gulf states were under less pressure to change, except perhaps with regard to improving the performance of the companies.

In general, the NOCs have been heavily taxed or have had to surrender a substantial part of their income to the government. In Venezuela, for instance, the effective rate of income tax was set at 65.7%, minus a rebate

[261] A.G. Kemp, 'Petroleum Policy Issues in Developing Countries', *Energy Policy*, February 1992, p. 105.

[262] *Ibid.*, p. 109.

for new investments of up to 2% of taxable income. The tax is based on the export values determined for tax purposes by the National Executive.[263] These 'values' may exceed actual selling prices, and the Venezuelan Congress decided in 1976 that , from 1986, the export values would not be allowed to exceed actual prices by more than 20%. On top of income tax, PDVSA had to pay royalty and some other smaller taxes on land and imports. PDVSA's contribution to the state coffers was substantial: "Even in 1990, with oil prices $10 a barrel off the heady days before 1986, oil's contribution to the Venezuelan economy was to provide 86 per cent of foreign exchange earnings, 83 per cent of all tax revenue and 24 per cent of GDP."[264] In comparison to other OPEC countries, Venezuela's level of taxation appears to be rather high. In 1998, the Hamaca extra-heavy oil scheme, which is a joint venture of PDVSA, Texaco, Arco and Phillips was thwarted because of severe fiscal pressures on PDVSA, which resulted in its inability to meet equity costs. [265]

In Nigeria, some 75-80% of federally collected revenue is dependent on oil revenue. This is made up of petroleum profits tax (PPT; "of 85% reducing to 65.75% for up to five years until capital expenditures have been recovered"[266]), mining rent (through charges, duties, and royalty) and of the earnings of the National oil company (NNPC).[267] Nevertheless, Nigeria changed its policy towards those private companies that were previous majority stakeholders in the 1970s. The fragile state of the Nigerian economy in the early 1980s[268] and the opportunity the government saw of tapping into the international oil companies' investment funds, technology and marketing channels stimulated the government to start an incentive scheme to draw more private funds into the oil industry. With the aim of both boosting production levels and new investments, the new incentive agreements of 1986 allowed the companies guaranteed profits of

[263] L. Randall, *op. cit.*, p. 177.

[264] *Energy Economist*, December 1991, p. 16.

[265] *International Petroleum Finance*, January 1999, p. 11.

[266] Nigeria: a special Report, *Petroleum Economist*, March 1992, p. 21.

[267] The Economist Intelligence Unit, *Country profile Nigeria 1993-94*, p. 34.

[268] M. Gavin, 'Adjusting to a Terms of Trade Shock: Nigeria, 1972-88'. In *Policymaking in the Open Economy, Concepts and Case Studies in Economic Performance*, R. Dornbusch (ed.), Oxford University Press, 1993, pp. 172-219.

$2 a barrel.[269] In 1991, a new agreement was reached that shored up the slipped profit margins of the international oil companies. Under the terms of this agreement, companies that invest more than $1.50 a barrel in new exploration and development receive a profit margin of $2,50 a barrel; those investing less receive $2,30 a barrel.[270]

The situation is similar in Ecuador, CEPE's income tax payments form a substantial part of central government revenues. A contractor's permitted profit varies from 15% on production up to 10,000 b/d to 20% on production over 100,000 b/d.[271]

Although the tax regimes have been under pressure to change, governments have been reluctant to make sweeping reforms. Despite increasing competition, their dependence on short-term oil revenues and the inconclusive discussions on the commercialization or privatization of the NOCs have resulted in a state of inertia.

7.4 Core skills

With varying degrees of success, the NOCs have been able to develop their core skills in a relatively short period of time; that is, asset specificity in OPEC countries has clearly increased since nationalization. Since the NOCs acquired their core skills through nationalization, not by entrepreneurial process or innovation, they required the assistance of their respective governments to create their strategic core. The assertion of market power by OPEC and the ensuing deployment of oil windfalls in the oil industry helped the NOCs to develop more core skills and obtain certain advantages. However, the firms need to protect and/or maintain their core skills, and since their strategic core is developed in a dynamic environment, they also need to stimulate continuous renewal of those core skills. The extent to which they are successful in defending, maintaining and renewing core skills is a function of their R&D, and human resource development capabilities, and the organization's flexibility in adapting to

[269] The Economist Intelligence Unit, *Country Profile Nigeria 1989-90*, p. 26.

[270] The Economist Intelligence Unit, *op. cit.*, p. 25.

[271] G. Barrows, 'A Survey of Incentives in Recent Petroleum Contracts'. In N. Beredjick and T. Wälde (eds.), *Petroleum Investment Policies in Developing Countries*, Graham & Trotman, London, 1988, p. 234.

its new environment.[272] Proper development of the strategic core in this way will lead to an expansion of the boundaries of the firm, which may take the form of forward vertical integration, backward vertical integration, horizontal acquisitions or diversification.[273] It is also possible to generate economies from strategic alliances.

Most NOCs have had trouble protecting and maintaining their (recently acquired) core skills. Moreover, NOCs that are tightly controlled by their governments and are in some way limited in their scope or functioning find it particularly hard to adapt to the changing environment of the firm. Management does not have complete freedom to invest and to develop a market strategy, nor is it free to define, let alone change, the efficient boundaries of the firm.[274] Furthermore, the NOCs in certain developing countries were functioning in a highly protected business environment, in which they had first choice in permits and no internal competition, and in which financial requirements were initially met on a needs basis. Moreover, they did not have to compete for crude reserves abroad, because their purpose was to supply the domestic market. Only when oil prices fell in 1986 and competition for finance increased with other sectors of the economy, the pressure on these NOCs to improve their efficiency increased.

In the 1980s, the national companies of Venezuela, Kuwait and Saudi Arabia, not only increased their skills in the various parts of the value chain in the domestic market, but they also acquired operations abroad. PDVSA and Saudi Aramco have also established joint ventures, and strategic alliances in consumer markets. KPC has developed into an international vertically integrated firm, pursuing an independent market strategy. In most countries, the oil industries are organized as vertically integrated firms. In general, NOCs that were able to integrate the nationalized assets of international oil companies fairly quickly, were able to develop satisfactory management structures and skills within the parameters set by the government. However, we must take note of the remarks of Shleifer and Vishny "When managers control firms, politicians use subsidies and

[272] T. Reve, *op. cit.*, in: M.Aoki et al., *The Firm as a Nexus of Treaties*, 1990, pp. 145-6.

[273] P. Milgrom and J. Roberts, *op. cit.*, 1992, pp. 538-83.

[274] B. Holmstrom, 'The Firm as a Subeconomy', *Journal of Law, Economics & Organization*, vol. 15, no. 1, April 1999, pp. 95-9.

bribes to convince them to pursue political objectives. When politicians control firms, managers use bribes to convince them not to push firms to pursue political objectives."[275]

In order to understand the environment in which national oil companies had to develop their core skills, it is important to consider the market cycle and the organizational structure of the oil industry in the 1970s and 1980s. In this context, it is imperative to examine the market strategies of the national oil companies' main competitors, the private international oil companies.

7.5 COMPETITIVENESS

At the beginning of the 1970s, the NOCs had little experience in R&D, and although increased spending on education substantially improved the supply of skilled labour, the demand for technological input was both too wide ranging and too specialized for the NOCs to develop it all at once. In fact, the NOCs were lagging behind the international oil companies in nearly all parts of the value chain. It is therefore understandable that the NOCs did not try to tackle everything at once, but selected certain priority areas within R&D which were of particular importance to the circumstances their company and country.

The Orimulsion project of Venezuela is one such project, in that it took the chemistry of the crude reserve base of PDVSA into account and was considered a rational project for PDVSA to develop in order to secure the future of the oil industry in the longer term.[276] Due to its declining medium and light crude oil reserves, PDVSA will produce more heavy crudes in the future. The problem was, however, that the Orimulsion project required a large amount of dedicated investment in specially equiped refineries and consequently also long-term supply contracts or joint venture constructions to reduce risks. Although, Orimulsion would not be competitive with lighter crudes in the short term, the project ensured that the Orinoco Bitumen reserves can be transformed into a marketable product in which PDVSA's technology is key. Also, PDVSA hoped that the fact that the company would be absorbing the initial R&D costs, would attract investors into the project who would be willing to swap refinery capacity for

[275] A. Shleifer and R. Vishny, *op. cit.*, 1994, p. 40.

[276] F. McGowan, 'The Development of Orimulsion and Venezuelan Oil Strategy', *Energy Policy*, March 1992, pp. 913-25.

upstream equity. Both PDVSA and private oil companies have something to offer: one has crude, the other has processing facilities and markets. Once sufficient dedicated investments are in place, the product's market will be secured. Casas-Gonzalez of PDVSA stated: "However, many companies who own such refineries are finding it more difficult and expensive to find and maintain upstream crude oil reserves -so this is obviously the key to our new strategic goals. In case of the Venezuelan heavy and extra-heavy oils, we think that there is a possibility to form new associations in the downstream business with oil companies who want to move back into upstream ventures in Venezuela. These could include the reactivation of oil fields closed long ago due to difficulties which have now been overcome by the technology and know-how that some companies have been perfecting in recent years, and the opening of frontier areas to oil exploration. These ventures could result in long term relationships of mutual benefit, especially since such long term, highly integrated joint ventures may enjoy an easier leverage of bank financing, as most uncertainties can be removed."[277]

The protracted period of excess production capacity in OPEC countries has been a disincentive to develop exploration skills. Considering their production levels and potential, fairly little exploration and development took place in OPEC countries. The technological advantages of private international oil companies with their 3D-seismic technology and new drilling techniques are immense. Contracting firms to conduct these services has become more costly. Not only are the international oil companies now more reluctant to sell their technology, but it also has become more expensive to NOCs since margins on oil production have shrunk. Many OPEC countries are now having to increase production capacity to ensure an adequate supply of oil in the future. Increasingly, this new oil has to be extracted from more difficult geological structures, is found in smaller oil wells and is more expensive to produce. The international oil companies have developed the skills necessary to produce this oil at the lowest cost and, aware of their technical competitive advantage, they want to exchange their skills for equity oil. The straitened financial position of governments, which reduces the availability of risk capital to the NOCs, has compelled them to allow foreign private investors back into the domestic oil industry. The fact that the international oil companies have managed to renew their

[277] A. Casas-Gonzalez, *op. cit.*, p. 7.

competitive edge is confirmed by Bourgeois and Rodriquez-Padilla: "(...) the OPEC countries and companies have been unable to reap the benefits of their enormous comparative advantage in hydrocarbon geology, and have been placed in a defensive situation where they are forced to slow or abandon their efforts to dominate the industry. To explain this reversal, we must take into account comparative advantages other than geological ones in analyzing the competitive situation. In this case, that means the advantage for a given player of market power in the downstream market and its overall ability to adapt to a new environment. Western companies have demonstrated a comparative advantage in these areas that at least partially offsets their geological disadvantage."[278]

Companies like PDVSA have welcomed the opportunity to engage in joint ventures with the large international oil companies because they consider the exchange of equity crude for technology, capital and markets beneficial. From the NOC's point of view, such alliances will reduce governmental control over the firm, allowing management to become more independent of government policy, and will remove the restraints of OPEC policy on the firms. Also, they view such alliances as a way of guaranteeing the company's long-term survival. The austerity measures that governments were forced to take as a result of their debt problems also affected the investment plans of the national oil companies. In Algeria, Indonesia, Iran, Iraq, Nigeria and Venezuela, the debt-servicing problems of the government constituted a major constraint on investment.[279]

The performance of some NOCs has been mediocre at best, and many inefficiencies remained hidden as long as oil prices were relatively high. Some inefficiencies were forced on the companies by the government, and condoned as necessary for maintaining employment, satisfying labour unions or aiding capital formation in the domestic economy. The prolonged buyers' market has brought these inefficiencies to light, and the companies see in the financial weakness of the governments an opportunity to pursue company strategies as opposed to government economic strategies. The economic perils have been so great in some OPEC countries that either the oil regime was never totally closed to private foreign investors in the first

[278] B. Bourgeois and V. Rodriguez-Padilla, 'Oil Exploration and Production Between the 1980s and 1990s: Geological Advantage and International Petroleum Orders', *Energy Studies Review*, Vol. 3, no. 3, 1991, p. 246.

[279] *Ibid.*, p. 251.

place or some liberalization of the oil regimes has already taken place. A number of OPEC countries have merely reduced the activities of the international oil companies. Finon explains: "In fact, the multinationals had never really completely left the financially, organizationally and technologically less well-endowed countries. In such countries as Ecuador, the UAE, Gabon, Indonesia and Nigeria, they continued to benefit from PSCs; in 1989, the multinationals enjoyed direct access to 13.5% of OPEC production."[280]

Over the past couple of years, the oil regimes in these countries have been further liberalized, and governments are implementing a policy to attract more foreign investment. Now that more countries have started to liberalize their regimes, there is growing competition among oil-producing countries in offering the international oil companies beneficial fiscal and contractual terms and greater access to crude production. Algeria and Iran have relaxed their policies with regard to investments by the international oil companies. Since 1992, Iraq has been on the side lines waiting to re-enter the international oil market with a pro-private investor policy when UN sanctions are be lifted.[281] Indonesia, in the wake of the Asian financial crisis and the ensuing change in government, intends to introduce sweeping changes in the oil regime. The government of Indonesia has put before parliament new minerals legislation which will allow private investors to bid directly bid for oil exploration contracts, will reform the fiscal regime and will remove Pertamina as the government's agent. If this new legislation is passed, Pertamina will be converted into a private company.[282] Due to opposition to the new legislation, allegedly led by Pertamina, parliament has yet to consider these changes.[283] Despite these policy changes, drastic reforms to make the oil industry more competitive and investments less vulnerable to price volatility have yet to take place in the large oil-producing countries in the Middle East. Saudi Arabia, to maintain its market share, still prefers to rely on low oil-price policies and periodic

[280] D. Finon, 'The Prospects for a New International Petroleum Order', *Energy Studies Review*, Vol.3, no.3, 1991, p. 263.

[281] *Petroleum Economist*, January 1992, p. 9.

[282] Indonesia, 'Draft Law of the Republic of Indonesia concerning Natural Oil and Gas', second draft 10 February 1999, Article 42.

[283] *Financial Times*, 13 July 1999.

production cuts to support prices within a \$14-\$18 price band. However, oil policy reforms have already been considered, and they seem to indicate that foreign direct investments in the oil sector, in order to increase competition, are being considered. The pressure on Saudi Arabia to liberalize the oil sector does not come with the same urgency as in other countries. The fact that the low extraction costs make the margin for inefficiency larger and that the government, so far, has been able to borrow on the domestic market makes the difference.

Further evidence of the ongoing liberalization and privatization trend is that Statoil, Norway's state oil company, has recently proposed the government to merge the interests with SDFI, since 1985 holder of 40% of Norway's oil and gas reserves, and become the 4th or 5th largest quoted company.[284] Norway's oil minister Langstein responded: "The government has an open mind for alternative proposals, but we need a broad political consensus. (...) It is much more than just the ownership relationship between the state and SDFI, and much more than the privatisation and stock market listing for Statoil -it is how we manage resources long term".[285] Although the political debate has by no means been decided and such a change has a large emotional charge, the pressure from the international market, and the advantages to the government budget in an alternative oil regime, might overcome the reluctance to privatize.

Now that the buyers' market has resulted in a large portion of the rent being shifted to the consumer countries, and the NOCs have failed to capture technical and managerial rents, the oil-producing countries and the NOCs find themselves at a crossroads. The NOCs need to concentrate on developing their core skills and securing the future of the firm in the international oil industry, while the governments need to ensure the continued flow of oil income stream required to manage the economy in the same way as before. The three-pronged strategy of OPEC countries (OPEC, national policy and companies) seems to have outlived its usefulness, and new strategies need to be developed.

The gloomy economic outlook for the economic performance of oil-producing countries following two successive oil windfalls should not tempt us into concluding that the policies pursued by the governments of

[284] *Financial Times*, 16 August 1999.

[285] *Financial Times*, 16 August 1999.

the OPEC countries have all been wrong. There is no denying the fact that the nationalization of the assets of the international oil companies in the 1970s has elevated NOCs in at least three OPEC countries to major vertically integrated players on the international oil market. However, as long as these companies remain in the public sector, the objectives of the NOCs will be mixed, and their performance will be unsatisfactory compared to their private sector rivals.

The attempts of governments in developing countries to reform (i.e., commercialize) the state companies rather than privatize them have been inadequate. The implementation of reform has been difficult, and in most cases the governments have failed to commit to hard budget constraints, particularly when the immediate crisis, which spawned these plans, has dissipated.[286] Horn concludes by saying: "Legislators decide what sort of objectives are embodied in SOE [State Owned Enterprise] legislation and the extent to which they subsequently intervene in the management of the enterprise. There is a remarkable regularity in these decisions: SOEs are typically established with mixed objectives and the government of the day frequently intervenes in management to impose noncommercial considerations. Moreover, this regularity appears to be very resistant to change."[287] In this regard, state ownership is shown to be less than optimal, since the varied goals of government policymakers often means that they will reallocate assets away from their most efficient use. It is worth pointing out that conflict of objectives is an unavoidable contracting problem, which, because of bounded rationality, makes it difficult to minimize this hazard. Any attempt to design alternative governance structures for NOCs must look to an organizational construction that provides for high-powered incentives for the parties to carry out the terms of the original. To be sure, the introduction of performance-based incentives for managers of NOCs, as a response to the defects in organizational design, might be a plausible technique to limit the diversion of assets for alternative inefficient uses. However, this strategy may not be sufficient as an incentive to change performance, making privatization the only realistic economic course of action.

[286] M. J. Horn, *op. cit.*, p. 137.

[287] *Ibid.*, p. 139.

7.6 Government interests and privatized NOCs

The concern of governments as to how they can secure their income stream from oil is understandable, particularly when they face severe debt problems. Governments therefore have a vested interest in the well-being of their domestic oil industry. The question is, then, how governments can best secure this income stream in the short and long term.

At the moment, OPEC countries do not have enough interests in common for there to be a full restoration of OPEC power for any prolonged period of time. The group is too heterogeneous to agree, national interests clearly prevail, and too much non-OPEC oil is still on the market. At national level, the budgetary problems of many OPEC governments are holding back any further autonomous development of the oil industry. If full privatization is politically unacceptable, investments in the oil industry should at least be supplemented by private foreign investments. To enable this, governments should allow private parties to invest and/or allow the NOCs to engage in joint ventures and/or offer some shares of the NOCs on the market to increase capital.

As an alternative to the old oil regime, in which income stream was secured through taxation of the state companies, it might be possible to use bidding systems for exploration acreage and some form of resource rent taxation as income-generating instruments, while the NOCs could be privatized. Particularly in the current situation, in which income from oil is relatively low compared to expenditure and the operations of some NOCs are not very efficient, the government, as far as its revenue is concerned, would be better off in a liberalized oil sector, despite the risk-sharing elements of the resource rent tax. In a new tax regime, capturing part of the economic rents requires conditional taxation, and taxes should be predominantly profit-related and flexible enough to account for the fluctuations in the size of the economic rents.[288]

The fiscal regime should take into account that private investors prefer to quickly amortize loans, possibly at an early stage in the project. After the loans have been repaid, a reasonable return on investment is necessary as an incentive for the investor. The application of a more flexible taxation regime that shares some of the risk with the investor will benefit the government through an increase in business, which will provide the state

[288] A.C. Kemp, *op. cit.*, *Energy Policy*, February 1992, p. 114.

with compensation for its reduced participation. Income streams can be preserved and sustained by fine-tuned taxation schemes.

The fiscal regime should not be a disincentive to investment, and the tax regime should be stable to allow the investor to calculate costs and benefits. Exploration, production and processing require long-term investments in dedicated assets. The government will need to avoid fiscal distortion effects on the economy. The United Nations Centre on Transnational Corporations has advised governments to implement a package of fiscal instruments which include, if a traditional or modern concession system exists, "a modest royalty plus a resource-rent tax and income tax (...). Where Governments of host countries prefer production-sharing agreements, an efficient package would be a combination of such a contract incorporating progressive scales whereby the State's share of "profit oil" increases with production plus a resource-rent tax. A system whereby the Government's share of "profit oil" increases with the rate of return attained can also readily be designed. This obviates the need to have a separate resource-tax payment."[289] The fiscal packages could vary according to the type of acreage and the risk involved. However, the fine-tuned application of fiscal instruments requires a sophisticated tax administration.[290] The administration would have to be adequately informed concerning the inputs or costs and the outputs, and be aware of oil industry procedures.

Governments that insist on a high level of participation in the NOC could find themselves acting against their own interests. A large commitment of capital to the oil sector when budgetary restrictions are in force might not be an appropriate investment, all the more so since, as regulators, governments can still exert considerable influence over the development of oil resources, without having to own a large stake in the NOC.[291]

The relationship between the NOCs and the state varies among the OPEC countries. Although state interference is considerable in all OPEC countries, the level of managements' independence varies substantially, as

[289] United Nations, United Nations Centre on Transnational Corporations, *Financial and Fiscal Aspects of Petroleum Exploitation*, UNCTC Advisory Studies, Series B, no. 3, ST/CTC/SER.B/3, New York, 1987, p. 33.

[290] T. Wälde, Investment Policies in the International petroleum Industry- responses to the current crisis, in: N. Beredjick et.al., *Petroleum Investment Policies in Developing Countries*, Graham & Trotman, London, 1988, p. 17.

[291] United Nations, UNCTC, *op. cit.*, 1987, p. 32 .

do the levels of commercialization, integration and internationalization. A process of commercialization is now under way in most OPEC countries and in some de-monopolization has also started. Complete privatization, though highly desirable, appears to be politically unfeasible in the near term. In many oil-producing countries, state control over oil exploitation is embedded in the constitution. What stands in the way of private sector exploitation of the oil resources, is the prevailing interpretation of state ownership; the state can only accomplish this control through a state oil company. However, the present practice of foreign direct investments suggests that contractual solutions can easily be found without changing the constitution. In the case of Pemex, sensitivities regarding the status of the company still run very high. In response to suggestions that shares of Pemex should be floated, Lajous, General Director of Pemex, stated that: "[p]rivatisation would inevitably hand Mexico's oil industry over to foreigners, because the domestic private sector industry is still in a fledgling state. That is a sacrifice he believes Mexico should not make: There are historical and constitutional reason's for Pemex's state control."[292] The introduction of a new regulatory framework, the introduction of competition in the sale of final products in which Pemex still has a monopoly, and strengthening of corporate governance should further the modernization of the Mexican oil sector. However, without a market for corporate control and only a partial opening to competition, the Mexican government might address the political problem, but not the economic one.

The size and sophistication of the private sector in a country is important for the success and political acceptance of privatization. The oil-producing countries still have a lot of work to do. Because of the inefficiencies in management and production, the companies will need to undergo considerable restructuring before any substantial share offerings can be contemplated.[293] Furthermore, the government will need to implement new business legislation and taxation schemes to attract private investors. The regulatory reforms necessary to create competitive markets are still far from being realized, and in some cases, governments may require IMF and World Bank assistance in their attempts. Liberalization and privatization

[292] *Financial Times*, 17 August 1999.

[293] J. Elting Treat, *National Oil Companies: Challenges for the Future*, Booz, Allen & Hamilton inc., Middle East Petroleum and Gas Conference, Dubai, January 13, 1993, pp. 26-7.

of the economy should be executed in the proper sequence in order not to replace one deficiency with another.

Some NOCs have been able to develop into vertically integrated international companies. In those cases, conflicts between the firm and the government is feasible or have already developed, as in the case of PDVSA's case, over the firm's strategies. Certain firm strategies can be perceived by the government as contrary to its economic policies. However, the government's public policies may, in turn, restrain the firm's progress. For instance, the rate of return on investments abroad may be more beneficial to the NOC than the rate of investment in the domestic economy, while the government attaches greater priority to capital formation and employment at home. In those cases where the NOC is able to compete on the international market and wishes to pursue an international strategy that the firm perceives as the most profitable, the government could contemplate a policy of non-interference. As a shareholder of the NOC, a profitable oil company will certainly benefit the government coffers more than a company that underperforms.

However, the lack of a market for corporate control does constitute a serious problem for the government as shareholder. Reducing the stake of the government in the NOC could create a sufficient basis for accountability of firm managers. Reducing its stake should be accompanied by the introduction of other regulatory reforms. First, the ownership of oil reserves should be de-monopolized and, second, a competitive bidding for exploration areas should be introduced. The introduction of competition in the oil sector is an important element in such reforms. It must be noted, though, that the literature on regulatory reforms is not positive about the efficiency of weak competitive regimes.[294]

The recent lack of economic stability has spoiled the OPEC countries' economic performance record.[295] However, this does not diminish the significance of the potential benefits of nationalization to OPEC countries. More than twenty years after the first oil-price increase, some of the NOCs are among the largest oil companies in the world, and some of these companies should be able to compete fairly effectively on the international markets when the liberalization process has been completed. One could say

[294] See Chapter 3.

[295] See Chapter 6.

that for those NOCs, the infant industry stage is over, and they should now be allowed to stand on their own feet.

CHAPTER 8

AMBITIOUS CONSOLIDATION

8.1 NEW COMPETITION

Technology and financial instruments have become very important in setting new investment standards. Technology, as it has been developed in exploration, production and processing since 1973, has served as a major cost-cutting device for international oil companies. Moreover, the new technologies have also provided the companies with alternative sources of crude, because deposits that were formerly uneconomic could be brought on stream. Particularly the developments in offshore and (semi-)arctic environments, such as the North Sea and Alaska, have turned out to be a large source of ingenious human capital and engineering innovations. The learning curve (through learning-by-doing) has been steep. These new technologies have created a new competitive edge for those international companies that have dared to venture into these high-risk locations. Moreover, it is exactly the willingness to take on this risk in unexplored areas (investment risk) under a volatile price regime (price risk) and the ability to manage these risks (joint ventures, technology, cooperation agreements, trading, hedging, etc.) that have changed the performance of private international oil companies. The rate of success of international oil companies in handling these various risks has also brought about a change in the structure of the international oil market.[296]

The sophistication of international banks in energy-related investment projects has maturated into much more risk-aware financing. The firm belief in OPEC's ability to stabilize the international oil market and to maintain the value of their oil reserves was shaken in 1985-1986. Moreover, the inability to appraise the economic instability as a result of external shocks, (the 'Dutch Disease', construction booms and busts and corruption) forced the banks to restructure many of their loans to oil-producing countries. The development of new financial products that covered the different types of risk that are involved in energy financing,

[296] J. Roeber, 'Oil Industry Structure and Evolving Markets', *The Energy Journal*, Special Issue, 1994, p. 274.

offered the banks more security.[297] The hedging of risks became a new standard in energy investment and introduced private foreign oil companies back into oil-related projects in countries where they had had no access since 1973. In addition to risk management, cost management increased in importance after 1986, when OPEC could no longer sustain output restraints that stabilized prices at a relatively high level. Competition among the oil companies increased, as did competition between private and public oil companies. In the 1980s, first- and second-tier international oil companies had taken over smaller companies and/or acquired crude reserves through takeovers and mergers, while, at the same time, the first- and second-tier companies had divested refining capacity in the mature consumer markets, and upgraded their remaining capacity. New refining capacity was predominantly installed in emerging markets in Asia.

The break-up of the value chain as a result of the nationalization of OPEC crude reserves and production resulted in two different company strategies in the period 1973-1986. The large international oil companies, predominantly the first-tier group (Exxon, Shell, BP) continued to invest internationally. Other companies consolidated their assets in certain markets, such as Europe or the US (Chevron, Mobil, Texaco). Although both groups of companies engaged in mergers and takeovers in the 1980s (particularly the second half of the 1980s), they must be explained differently. Non-value-maximizing mergers and takeovers were much more frequent in the 1980s than in the 1990s.[298] Oil companies had a lot of free cash flow that they needed to invested or to diversify their portfolio. In the absence of new exploration areas some companies simply bought other companies' crude oil reserves and production (e.g., Texaco/Getty Oil; Chevron/Gulf Oil). The free cash flow at that time also resulted in some share buy backs that were justified as a protection against takeovers (e.g., Shell bought back the remaining shares of Shell Oil in the US, of which it was already a large shareholder). The acquisition of BP of the remaining shares of Sohio to create BP America must be seen as a value-maximizing merger in which operational efficiencies could be gained in the aftermath

[297] G. Pollio, 'Financing Energy Development In An Uncertain World, The Role Of Project Finance', Presentation, Faculty of Law, Leiden University, Spring 1993, Exhibits 32-34.

[298] R. Romano, 'A Guide to Takeovers: Theory, Evidence, and Regulation', *The Yale Journal on Regulation*, vol. 9, 1992, p. 121.

of OPEC nationalization and BP's market position in the US could be solidified. The acquisition of crude oil assets by non-oil companies and merger and acquisitions among smaller independent oil companies reflected their financial and technical inability to engage in frontier developments.

In the late 1980s and early 1990s, when the returns on investment in the US were declining as compared to the situation abroad, the strategy of the oil companies to consolidate their position became, once again, more internationally oriented. Competition for new crude reserve and production facilities increased when the oil deposits in the Soviet Union became available after the break-up of the state. In order to compete successfully and manage risks and costs, the first-tier international oil companies consolidated their position through large mergers with and takeovers of second-tier oil companies. This type of mergers, unlike most mergers and takeovers in the 1980s, falls largely in the category of mergers for value-maximizing purposes.[299] Value-maximizing explanations can be subdivided into: (1) Value-maximizing efficiency explanations, in which the acquiring company seeks synergy gains, such as operational efficiencies and financial synergies, and seeks to reduce agency costs, by replacing inefficient management and creating free cash flow; (2) value-maximizing expropriation explanations, which focus on tax, bondholders, employees and consumers; and (3) value-maximizing inefficiency explanations, such as underpricing and market myopia. The takeovers of Mobil by Exxon, Amoco and Arco by BP, and Petrofina by Total were announced as synergy mergers and takeovers.[300] However, such a judgement, particularly in the case of the Exxon/Mobil merger, can only be passed when the divestitures as a result of competition law rulings have been completed. Moreover, some consider the Arco takeover by BP Amoco as a non-value maximizing takeover, i.e., is a form of self-aggrandizement on the part of the management to truly belong to the "super-majors club".[301]

As a result of both privatization of state oil companies and the large mergers and takeovers, the PIW ranking of oil companies, in which the top

[299] *Ibid.*, pp. 122-45.

[300] *Financial Times*, 28 November 1998.

[301] 'What BP Amoco sees in Arco', *Energy Intelligence Group*, March 29, 1999. (Http://www.piwpubs.com/U990329A.htm)

10 was dominated for two decades by the state oil companies of Saudi Arabia, Venezuela, Iran, Indonesia, Mexico and Kuwait, is changing rapidly in composition. Private international oil companies are increasingly catching up on the largest state oil companies, and a super 'league' of large international oil companies, competing for dominance, is coming into existence again.[302] This might add further increase pressure on state oil companies to do away with the inefficiencies with which they are usually correlated.

In dynamic markets, competitive advantages, delays and irregularities in developments generate differences in the market positions of companies.[303] In the 1990s, the combination of the privatization of state oil companies and the mergers and takeovers among the first- and second-tier private international oil companies has stimulated a dynamic process of jockeying for market positions and created competition among giants. At the same time, this process of intensified competition during a period of lower oil prices will entice an impulse to consolidate and will result in market concentration.[304] The understanding of the present reorganization of the international oil market is incomplete without an explanation of the responses of international private oil companies to nationalization and OPEC market power.

8.2 THE INTERNATIONAL OIL COMPANIES

The nationalization of the oil reserves of international oil companies drastically changed the risk profile of these companies. Before, large oil reserves and/or term contracts guaranteed security of supply. In addition, particularly under the price administration of the 'Seven Sisters', the equity crude provided the companies with a source of rent appropriation and tax benefits under the tax neutrality acts of many oil-consumer countries. The loss of equity oil of international oil companies to the OPEC countries implied a rent transfer that radically changed the revenue streams of inter-

[302] 'Privatization Changing the Face of the World Oil Industry According to Ranking the World's Top Oil Companies', *PIW*, Energy Intelligence Group, June 10, 1998 and December 1998. (http://piwpubs.com/prpiwtop.htm)

[303] H.W. de Jong, 'Concurrentie en Concentratie in de Markteconomie'. In J.G.C. M. Bune, J.G. Groeneveld et al. (eds.), *Actualiteiten en Accountancy '91*, Nivra, Samsom Bedrijfsinformatie, Alphen aan den Rijn, 1991, p. 188.

[304] *Ibid.*, p. 189.

national oil companies.[305] Moreover, crude oil prices increased around 400% in a period of 9 months in 1973-1974. This impacted the downstream industry, which became the main activity of the international oil companies as a result of the enforced backward de-integration. Revenues had to be generated in transport, refining and distribution, and later trading, which changed the emphasis towards the higher-value-added part of the value chain, or the companies had to invest in new equity oil to maintain revenues from the entire value chain. The international oil companies began to invest in both OPEC and non-OPEC countries, provided they offered sufficiently advantageous fiscal and contractual terms and access to participation crude.[306] The oil price increase also made it possible to invest in frontier areas like Alaska and the North Sea. At the time it was believed that these companies were seeking full vertical integration again. However, the trade and investment pattern, as it evolved in the late 1970s and 1980s, proved otherwise, and full vertical integration was never achieved nor aspired.

The market power of OPEC member states in the 1970s replaced the oligopoly of the international oil companies and led to some major changes for the private international oil companies:

- excess equity production over processing capacity of the earlier decades belonged to the past;
- their main activities shifted, in a short space of time, from production to processing and distribution;
- their main activities were geographically relocated to traditional consumer markets, where the bulk of the processing capacity was established;
- the sea transport sector needed restructuring;

[305] The oil price increase initially led to huge income streams for international oil companies, because the nationalization took about 3 to 4 years to realize. However, from 1973 onwards the international oil companies knew that revenue streams from their OPEC crude oil production would be reduced within the next few years. Some OPEC countries did not fully nationalize but allowed the international oil companies minority participation in their oil production. Others realized full nationalization by 1980. In terms of structural income streams, the international oil companies knew already in 1973 that they would have to generate revenue elsewhere or compensate otherwise for their OPEC assets.

[306] D. Finon, *op. cit.*, *Energy Studies Review*, vol.3, no.3, 1991, p. 263.

- the processing sector, as a result of higher prices, needed to be restructured to comply with new supply and demand patterns;
- the former taxation of posted prices (the golden gimmick) no longer provided the companies with large after-tax cash funds; and
- trading, internalized for so long, became a new activity and profit centre. Some crude rich companies had to newly set up these departments.

The implementation of the nationalization schemes at the end of the 1970s ended a transition period for the international oil companies. From 1973 until 1979-1980, they had handled the crude for the OPEC governments and their state oil companies. After the 1979-1980 price increase, OPEC governments with nationalized industries took full control and diversified their customer portfolio.[307] From then on, the international oil companies had to either buy oil on the term market or the spot market, or invest in new equity crude to supply their downstream operations. The price volatility in the post-1973 period, and particularly after 1980, had turned term contracts into a source of risk rather than a security.[308] The trading strategy of companies had to incorporate price risks. To this end, new trading products were created, and hedging instruments were actively used. Price volatility reduced the duration of trading contracts and a relatively large amount of oil was traded on spot and forward markets.[309] Although trade in oil (volumes of oil moving) decreased in the period from 1973 to 1986,[310] both the wet market (spot crude) and the paper market (futures, options, etc.) were expanding rapidly, and matured into sophisticated markets, not unlike the financial markets that developed in the same period. The expansion of these markets also attracted new market participants who had previously stayed outside the realm of the oil market. In addition to the trading departments of international oil companies,

[307] R. Mabro, 'The Changing Nature of the Oil Market and OPEC Policies', *OPEC Review*, vol. VI, no. 4, Winter 1982, p. 325.

[308] Joe Roeber, *op.cit.*, p. 259.

[309] John Evans, *OPEC, Its Member States and the World Energy Market*, A Keesing Reference Publication, Longman, Harlow, 1986, pp. 93-6.

[310] J. E. Harthorn, *op. cit.*, 1993, p. 37.

arbitrators, banks and speculators participated actively in the paper markets and made the markets more liquid.[311]

The growing weakness of the OPEC to administer prices was an impetus for this new trading regime. In the mid-1980s, the oil prices moved from administered prices to market-based prices. The price fall of 1986 reduced the ability of governments or their state oil companies (depending on their legal structure) to invest in capacity additions and/or maintenance. State oil companies (successfully) tried to borrow on international capital markets to satisfy their investment needs. Since the investments were linked to current oil prices[312], the inability of OPEC to stabilize price increased price volatility and investment risks.

The nationalizations had greatly upset balance between upstream and downstream activities of the major oil companies. The forced process of backward de-integration radically changed the market position of the companies. Their main activities had shifted to the downstream part of the value chain, and their geographical scope had been narrowed temporarily. In 1987, the share of US production in the worldwide production of the larger international oil companies was around 50%.[313] Only Occidental, Exxon and Shell produced a larger share of their world production outside the US. Oil production of BP was predominantly concentrated in the US and the North Sea. Only in the late 1980s did many companies return to more international exploration and production activities, particularly when the former Soviet Union opened up for investments. Exxon and Shell maintained their international approach to the market throughout.

The refining industry was particularly affected by the crude price increases in 1973-74 and 1979-80. Furthermore, their refineries were outfitted to process mainly the lighter crudes, which gave them the best yield on the refined product mix. With the consumer governments moving their electricity generation away from fuel oil, the chemistry of demand was moving towards the whiter end of the barrel. The problem was, however, that crude supply in the 1970s was becoming heavier. The heavier crude

[311] P. Horsnell and R. Mabro, *Oil Markets and Prices, The Brent Market and the formation of world oil Prices*, Oxford University Press, Oxford, 1993, pp. 104-5.

[312] P.-H. Bourrelier, X. Boy de la Tour and J-J Lacour, 'Energy in the Long Term, Mobilization or laissez-faire?', *Energy Policy*, March 1992, p. 206.

[313] C. van der Linde, *op. cit.*, 1991, p. 198.

input combined with the equipment installed in the refineries resulted in output of oil products that was heavier than the market demanded. This confronted the international oil companies with another imbalance in their operations. The oil companies resolutely restructured their refining sector, they made the refineries more flexible to handle different types of crude and adapted those refineries that allowed crude switching. Refining capacity was substantially reduced and the remaining capacity upgraded.[314] Moreover, Mobil and Shell decided to invest in downstream joint ventures in Saudi Arabia and export products from these refineries.

As part of their new exploration and production strategy, the international oil companies increased investments in the newly discovered oil fields in the North Sea and Alaska, where costs of production were much higher than in the Middle East. However, the price policy of the OPEC safeguarded prices at a level that made oil produced in these areas profitable. Moreover, the development of fields in the harsh environment of Alaska and the North Sea, meant that the international oil companies accumulated a tremendous amount of experience and technological advantages. As a result of this learning process, the initial high production, finding and development costs in the North Sea were reduced in 20 years time to about $11 a barrel today. Compared to the $2 per barrel a day in the Gulf, the North Sea oil is still fairly costly to extract, but compared to extraction costs in North America ($10-11), the North Sea has become competitive. The extraction costs in Russia ($14) make the area the most uncompetitive.[315] The new techniques gave the international oil companies a competitive advantage in the exploration of harsh environments and offshore.

In terms of their core skills, the large international oil companies lost important site-specific and dedicated assets, and the boundaries of the firm were involuntarily redefined. Although the international oil companies immediately stepped up exploration and production in 'investment-safe' areas like the North Sea and Alaska, these investments could not effectively replace their former vast rent- and profit-generating activities in OPEC countries. Rather, the production of higher-cost oil compelled the

[314] *Ibid.*, pp. 193-5.

[315] 'Cheap Oil, The next shock?', *The Economist*, 6 March 1999, p. 23.

oil companies to seek alternative ways to generate rents and profits, and to regain their competitive advantage through R&D and innovation.

8.3 Investment Risks

In the same period that the oil industry's risk profile changed, the international capital markets came to appreciate the sources of risks, too.[316] The post-1973 'more-or-less' flexible exchange rate regime and the international debt crisis that dominated much of the 1980s made investors and banks more aware of the various pitfalls of international lending and investment.[317] In addition to political and economic risks, the investment banks also ran credit, market and transfer risks. The volatility of oil prices created a continuous appreciation and depreciation of the collateral, i.e., the crude oil reserves. In the early 1980s, it was not always possible to properly assess the risks involved. Countries like Venezuela, Mexico, Algeria and Nigeria, who were capital-short despite the expansion of their oil revenue, encountered debt-servicing problems as a result of widely fluctuating oil prices and oil revenues.

Confidence in OPEC's ability to regulate the international market began to erode after 1982 and this affected the investment scope of state oil companies when OPEC was obliged to continuously reduce its production quota. At first, this was not a particularly serious problem because the capacity utilization rate of OPEC countries was declining. But from 1986 onwards, as oil prices fell, this rate increased. The new low oil price became an impediment to investments. For the state oil companies, the fact that they had not been able to create a stable relationship with international banks became a problem. Also, in many oil-producing countries, the government, the ministry of finance or, as in Indonesia, the central bank was the main decision-maker on investment and the state oil company was merely an operator. In other countries, the state oil company had more

[316] M. Brennan and E. Schwarz, 'A New Approach to Evaluating Natural Resource Investments', and D. Siegel, J. Smith and J. Paddock, 'Valuing Offshore Oil Properties with Option Pricing Models'. In D.H. Chew jr., *The New Corporate Finance, Where Theory Meets Practice*, McGraw-Hill, New York, 1993, pp. 98-107 and 108-16.

[317] C.W. Smith, C.W. Smithson, D.S. Wilford, 'Managing Financial Risk'. In D.H. Chew jr., *op. cit.*, pp. 370-91.

freedom to decide over investments and in the way in which they were financed.

In Venezuela, this led to a situation where the state oil company PDVSA initiated an internationalization process, that was not entirely supported by the government. PDVSA invested in downstream facilities in the US and Europe and, in general, was able to secure international finance for these ventures. At the end of the 1980s, the credit-rating of PDVSA was superior to the credit-rating of the government. This created tension between the ministry of finance and the state oil company because additional borrowing by the oil company worsened public debt. The government began to resist the internationalization plans of PDVSA, actively curbing its corporate strategy and demanding that borrowing should serve domestic investments rather than international ones.[318]

Oil produced outside the realm of OPEC was attractive from an investor's point of view because, in the early 1980s, OPEC was prepared to mitigate price fluctuations. The investment risk was thus cushioned, and allowed the international private oil companies to venture into technically risky projects. When OPEC could no longer provide this safety net for investors, the cost of producing this marginal oil had already decreased substantially. Although the cost of producing of this oil was higher than for onshore oil in OPEC countries, oil prices never dropped far enough to prevent the private international oil companies from reaping a handsome profit -that is, until 1998-1999, when oil prices slipped to $10-11 a barrel for a 9-month period. Only when OPEC member states and Mexico agreed to restrict oil production in March 1999 did prices recover again.

This competition from non-OPEC oil continued even after oil prices crashed in 1986. Income from oil, already declining since 1982 as a result of volume reduction, decreased particularly strongly in those countries that were producing at or close to capacity. They did not have the power to raise income through sales of larger volumes unless major investments were made to increase production capacity. Governments were under severe budgetary pressures due to the decline in oil income and could not afford to make these investments. For many oil-producing governments, the only option was to allow their state oil companies to borrow abroad, which might worsen the credit-rating of the government, or to allow the privatization of (all or part of) the oil industry. The enigmatic and/or un-

[318] L. Randall, pp. 161-9.

proven record of some state oil companies created a situation that made borrowing on the international capital markets complicated.

International banks required more security after 1986, when only the variable interest rate and percentage points above LIBOR had incorporated risk. Despite their lower earnings since the early 1980s, the credit rating of most state oil companies was still fairly good, but the rating of most oil-producing governments was decidedly less so. International banks particularly requested guarantees against unwarranted intervention in the state oil company's investment plans by the sole shareholder, the state. In the 1980s, project finance regained importance in financing investments in the oil and gas industry.[319] The banks wanted to identify the investment object and make sure that sufficient cash flow could be generated from the project to reduce the credit risk. Moreover, they required projects to be sufficiently hedged for price and currency risks.

In addition, banks were concerned about the costs of exploration and production, which would ultimately translate into a profitable venture. Due to OPEC's production policy, most countries either had, not been forced to actively explore for oil themselves or, in the past, had contracted foreign companies to do it for them. The level of know-how in the state oil companies about exploring for and producing oil, particularly in the more difficult areas, had kept pace with the levels of know-how in foreign companies that had been actively involved in exploration in the 1970s and 1980s. In particular, 3-D seismic research, horizontal drilling, floating oil rigs, enhanced recovery, and so on have led to a drop in production costs and have increased the efficiency of existing oil wells.[320] With few exceptions such as the Saudi Arabian fields near Riyadh, the new exploration areas were in relatively difficult geological or geographic locations. Many of the recent additions to oil reserves have been found offshore rather than onshore, and therefore require more capital and more advanced technology.

The ownership structure of the company and the difficult economic situation in which governments in countries founds themselves were a source of possible conflicts of interest between the oil company and the government because in principle the government could instruct the

[319] J.W. Kensinger and J.D. Martin, 'Project Finance: Raising Money the Old-Fashioned way'. In D.H. Chew jr., *op. cit.*, pp. 323-35.

[320] J.E. Hartshorn, *op. cit.*, 1993, pp. 247-8.

company to use borrowed funds for purposes other than those for which they were originally intended.[321] Banks could secure against this eventuality by specifying the purpose for which the loan was to be used or by providing project finance. Project finance has been a widely used method of financing large-scale energy-related projects. The inclusion of a foreign partner alongside the state oil company provided banks with additional security and also ensured the input of foreign capital and technology.[322] The inability of governments to provide their companies with investment capital and the inability or unwillingness to put public funds at risk encouraged the development of this new structure for financing projects.

The use of project finance, in effect, started off a gradual process of privatization process in oil-producing countries, since it meant that foreign companies were able to secure investment projects in these countries that had been unavailable to them since 1973. This process of privatization due to economic necessity was further stimulated by the opening up of the market in the former Soviet Union. Competition for capital among oil-producing countries became much stronger in the early 1990s and opened up new markets to international oil companies. As a result of the collapse of OPEC's and the opening up of the former Soviet Union, competition in the international oil industry became unprecedentedly fierce. The reorientations that the international oil companies underwent in the 1970s and 1980s towards processing at the higher value-added end of the barrel and towards producing oil in frontier areas became their strongest assets. Those companies that, after nationalization, had actively sought to develop new skills rather than consolidate were among the first to move into the former Soviet Union. They had acquired technology and risk-management skills that the state oil companies and the less adventurous private oil companies lacked. When the opportunities arose, the international oil companies were much better positioned than the state oil companies to compete in this new market for crude oil reserves and production.

[321] The explosion of a badly maintained gas pipeline in Mexico City in the late 1980s led to PEMEX being instructed to give investment in maintenance priority. The company had just obtained a loan from international banks to invest in new exploration and production, and had no funds to carry out the maintenance. Due to the banks' serious threat to recall the loan, they were able to prevent these funds from being redirected to this non-income raising activity.

[322] G. Pollio, *op. cit.*, Exhibits 32-34.

8.4 From Disadvantage to Competitive Edge

As a result of the disintegration of the value chain in the 1970s, the international oil companies had to update their core skills, develop new competitive advantages and create new boundaries of the firm. This development also occurred in other industrial sectors.[323] There, the production of raw materials and some simple processing industries were taken over by developing countries' local producers or state companies. Large international companies then moved up in the value chain (dealing with design, marketing, distribution, and new products or materials), with production and bulk processing being left to smaller companies or state companies in the producing country. This process of industrialization in the developing countries normally took place when production and simple manufacturing had matured (as in the case of cotton and textiles, for instance). At that stage, consequently, competition in those markets was no longer based on quality but on price competition, making economic rents easy to capture at the higher end of the value chain.[324] International trade in price-competitive sectors such as cotton and textiles were heavily regulated under GATT, while in the quality-competitive sectors, self-regulation at company level could be employed. The multinational companies (MNCs) in price-competitive sectors thus effectively maintained a buyers' market.

The situation on the international oil market was different. A cocktail of political turmoil in the Middle East (October War 1973) and the particular oil market conditions prevailing in the early 1970s gave the OPEC countries cartel power.[325] In this crude oil sellers' market, the economic rents were captured by the oil-producing countries and/or their state oil companies, as the seven large international oil companies (Exxon, Shell, BP, Mobil, Gulf, Texaco, Chevron) had done before them in earlier times.

The strategy of the international companies to secure and control supply of raw material (oil), as they had done before 1973 through vertical integration, could no longer be implemented because nationalization effectively prevented them from accessing the bulk and the cheapest oil

[323] W. de Haan, F. van der Molen and C. van der Linde, *op. cit.*, pp. 123-67.

[324] D. Tussie, *The Less Developed Countries and the World Trading System: A challenge to the GATT*, 1989, London/New York, Pinter Publishers.

[325] C. van der Linde, 'OPEC Adrift', *Resources Policy 1994 20 (2)*, p. 96.

reserves. Investments in marginal oil fields entailed both a huge investment risk and also implied that, in order to survive, huge cost reductions would have had to be achieved in a fairly short period of time, i.e., the duration of OPEC's cartel power.[326]

As long as the largest oil-producing countries (the Gulf States) had only limited absorption capacity they could afford to support the high oil-price policy favoured by the smaller oil-producing countries in OPEC after 1980. The window of opportunity for the international oil companies to engage in risky exploration ventures was small, and could only last as long as OPEC was prepared to stabilize prices on the international market at a fairly high level. However, in the long run, the lack of homogeneity within OPEC would make the organization weak and would gradually erode its ability to reach internal agreement on policies.[327] Moreover, the investments made by international oil companies in new oil fields outside the realm of OPEC effectively undermined the continuation of the sellers' market. This same sellers' market provided international oil companies with protection for their high-risk and costly investments in crude oil production in the short run but at the same time undermined the profitability of their downstream activities.

It is obvious that the international oil companies were caught in the worst possible situation, with the loss of their low-cost oil activities in OPEC countries endangering the entire revenue-generating capacity of the firm. Not surprisingly, in the 1980s quite a few private oil companies (e.g., Gulf, Getty Oil) disappeared as a result of bankruptcy or near bankruptcy, merger and takeovers. Other companies engaged in a process of downstream disinvestment, rationalization and de-internationalization, effectively withdrawing to the 'safe' markets of North America and Europe. This shake-out of oil companies particularly affected those that had been crude long and had relatively few or small-scale geographically diversified downstream industries.[328] Crude long companies usually depended on oil

[326] H.W. de Jong, *op. cit*, 1985, pp. 151-64. Cartels have by definition a limited life span.

[327] Coby van der Linde, *op. cit.* , 1994, pp. 95-100.

[328] For example, after the nationalization of their oil assets, the fortunes of the old Aramco partners (Exxon, Texaco, Mobil and Chevron) were very different. Mobil was deeply vertically integrated, but geographically not very diversified. Texaco had its main downstream operations in the US and had been crude-long because

production from a few very large sources and as a result, did not actively undertake oil exploration for a considerable period of time, because they could not or were not willing to expand at the same pace downstream. The Saudi Arabian oil fields, for instance, were either discovered in the 1930s and were fully developed after 1945. Subsequently, no active exploration was undertaken in frontier areas and this had reduced or impoverished these crude-long companies' in-house human capital in exploration. In that sense, the initially advantageous size of the oil assets in countries like Saudi Arabia turned out, in the longer term, to be debilitating.

Those companies with shallow integration could, in the 1970s, not easily redirect their activities towards exploration and processing and therefore relied heavily on continued deliveries from OPEC countries. As long as OPEC countries did not have the distribution channels through which to sell their oil, these companies prospered. But following the second oil-price increase in 1979-1980, the OPEC countries completed their process of nationalization and found alternative distribution channels. The traditional sellers of this oil, the international oil compaies, were replaced and were left with only few sources of oil, mostly in the US. The changing market structure and their lack of foresight made them vulnerable to takeovers. Their share price was depressed and, as a result, proven reserves on mainland USA could be purchased on Wall Street (at approx. \$4-\$5 a barrel) far below the development costs of new oil wells (approximately \$18-\$20 a barrel in the North Sea). In the middle of a more general takeover boom in the US, which was fuelled by the enormous capitals that could be raised on the junk bond market and the tax treatment of corporate debt, a massive restructuring of the oil industry took place. Shell and Exxon managed to consolidate their positions as the largest international oil companies, and small or even tiny oil companies engaged in hostile takeovers of the larger companies. The attack on Gulf by Boone Pickens drove the company into the arms of 'white knight' Chevron, while Texaco

of its holdings in Saudi Arabia. Exxon was both deeply vertically integrated and geographically diversified. Gulf Oil, the 50% partner of BP in Kuwait, was also crude long, and sold its surplus crude oil in long-term contracts. Gulf had few activities in petrochemicals and predominantly sold gasoline. Gulf was not deeply vertically integrated.

snatched Getty Oil away from Pennzoil at the expense of near-bankrupting legal costs and huge damage payments to Pennzoil.[329]

The more challenging companies and/or the largest ones took a different approach. They did not reduce their international reach, but instead tried to rationalize their activities in processing, transport and distribution in the more mature consumer markets in the US, Europe and Japan, at the same time embarking on greenfield refinery projects in the Middle East and Asia. To counter the loss of their equity crude, the international oil companies were willing to take on the risks involved in frontier exploration in addition to their trading activities. If they managed to produce this marginal oil and achieve progressive cost reductions, the investments would give them a tremendous advantage.

In order to reduce risk exposure, most new oil developments were organized as joint ventures. Interestingly, quite a few of these early joint ventures were between Exxon and Shell. In the North Sea, for instance, Exxon and Shell jointly invested in oil fields, while Shell was the operator of the joint venture. BP, however, persisted in its old strategy, and managed to stay fairly crude-rich as a result of very successful exploration activities. BP's downstream activities, however, remained weak. Initially, BP merged its distribution assets with Texaco, in a venture that bears similarity to code-sharing in the international airline sector. Later, Shell established downstream links with Texaco in the US market. In 1998-1999, BP surprised the international oil market with two take-over bids for US oil companies. First, it took over AMOCO, and, before the integration of AMOCO into BP was seriously underway, it also made a bid for ARCO. BP's downstream activities have always been relatively weak, and the company's strong position in crude oil exploration and production made it too vulnerable to oil-price fluctuations. This weakness could undermine the continuity of the company's extremely successful exploration activities, and could, ultimately, make it vulnerable to a takeover. The takeover of AMOCO, in particular and, to a lesser extent, that of ARCO, should be seen as an aggressive strategy to secure the company's long-term future.

[329] T. Petzinger jr., *Oil & Honor, The Texaco-Pennzoil Wars, inside the $11 billion Battle for Getty Oil*, G.P. Putnam & Sons, New York, 1987. See also R.H. Mnookin and R.B. Wilson, 'Rational Bargaining and Market Efficiency: Understanding Pennzoil v. Texaco', *Virginia Law Review*, vol. 75, 1989, pp. 295-331.

As a result of these takeovers, BP Amoco became the seventh largest oil company in the world in the PIW ranking of 1998.[330] In that same ranking, ARCO was listed 22nd in 1997.

In the new markets in Russia, Kazakhstan, Turkmenistan and other former Soviet States, new alliances have been formed both with local and/or regional companies and with large international oil companies. Typically, new investment projects have several partners in which one or two of them become the operator(s). These alliances reflect a new economic and political reality among the international oil companies. The amount of equity oil that they produce in this way has increased substantially, most of which is sold on international markets rather than domestic markets.

The international oil companies' ability to overcome the setbacks they suffered when OPEC countries took control of the market and to compete successfully in terms of capital and technology reduced the market power the oil-producing countries' large crude reserves and their ownership of production capacity. As early as 1982, the oil market became a buyers' market, frustrating OPEC's control over the market. The new additions to world crude reserves have so far predominantly been in non-OPEC countries, the cost pf producing and this oil is being such that it allows the international oil companies to compete with low-cost oil-producing countries. The oil-price fall of 1986 confirmed that, although the oil-producing countries can extract oil at very low cost levels, they cannot afford to compete at that price level. Governments' dependence on a certain level of income for their budgets has effectively increased the minimum price to the \$14-\$18 region, at which price private companies can easily produce oil from more marginal fields.[331] It is this price level that has drawn enough oil into the international market to enable the crude buyers' market to persist for such a long time. The ability of the international oil companies to produce marginal oil gives them substantial leverage over the market. Lowering the cost of the marginal barrel of oil has meant that there is more oil on the market capable of competing at the fringe with the low-cost oil from the Middle East. Such a situation makes it harder for OPEC or any other cartel of producers to increase prices

[330] *Company News*, 15 December 1998, PIW Study, Energy Intelligence Group, http://www.piwpubs.com/Topco.htm, p. 2.

[331] 'Cheap Oil', *The Economist*, 6 March 1999, p. 23.

beyond a certain level. Moreover, price volatility and the recent low price levels have created adjustment problems for oil-producing countries. This has driven the oil-producing countries into the arms of the international oil companies to seek new investments for their industry. Due to privatization as well as mergers and takeovers in the private oil sector, state oil companies are increasingly being challenged by private oil companies in terms of market capitalization.[332]

8.5 TECHNOLOGY AS A COMPETITIVE STRATEGY

The key element in the international oil companies' strategy has been their successful introduction of a new link in the value chain, one that precedes exploration: Research & Development. The scarcity of areas that were relatively easy to explore first led the companies to develop new technological solutions, such as 3-D seismic, horizontal drilling and floating and re-usable oil rigs. Then, in the 1970s and 1980s, the scarcity of cheap equity oil forced the companies to rationalize their downstream, transport and distribution operations. More recently, the upsurge in environmental legislation has continued the pressure on the R&D departments of international oil companies to come up with new solutions and technologies to produce oil at lower social cost.[333] Consumer governments in the OECD are increasingly taxing fuel according to its emissions of carbon dioxide, sulphur and lead, and determining the content of reformulated gasolines.[334] As a result, international oil companies have developed new core skills and are able to use their technological and managerial skills to compete with superficially low cost oil. The NOC's inability to produce this low-cost oil efficiently is due to capital costs and their inefficient way of dealing with managerial, social and technological matters compared to the large international oil companies. They are presently bad at converting 'low-extraction-cost oil' into 'low-production-cost oil', and, as a result, have lost a good part of their competitive advantage. The combination of their dependency on oil income with the present fairly high utilization

[332] *Company News*, 15 December 1998, PIW Study, Energy Intelligence Group, http://www.piwpubs.com/Topco.htm, p. 1.

[333] M. Williams, 'End of An Era', *Petroleum Economist*, September 1991, p. 41.

[334] *Petroleum Economist*, April 1992, p. 4.

rates has robbed them of their ability to compete on price and has reduced them to reliable bulk suppliers in the international market.

8.6 LIVING WITH LOW OIL PRICES

R&D has resulted in continuous pressure on the marginal barrel market in non-OPEC or non-state-dominated oil-producing countries. Extraction costs (except in the Middle East, where they are $2 a barrel) vary from an average of $6 a barrel in Indonesia and $7 in Nigeria and in Venezuela, to $10-$11 in Mexico, the US and Canada, and the North Sea. The highest extraction costs are found in Russia. In terms of competitive advantage and the main markets for crude, we can conclude that the state oil companies' competitive edge is not very substantial. Venezuela and Mexico export 61% and 89% of their oil exports, respectively to the US, where extraction costs are on average $3 higher.[335] Given the inefficiencies of the state oil companies, the profit margins must be much smaller. An inroad on these countries' market share from low -extraction-cost Middle East suppliers such as Saudi Arabia, Kuwait and Iraq (Iran and Lybia are still banned from the US market under the D'Amato legislation) is a credible threat. Due to increasing transportation costs, it is much more difficult for PDVSA and Pemex to compete efficiently on markets other than the US. The netback results will decline if more distant markets have to be found. Indonesia and Malaysia face the same problem in their traditional export markets, where Middle East producers could also pose a threat.

For non-Gulf producers, competition has been increasing both from the Middle East and from the private international oil companies, as they are wedged between the 'low-extraction-cost' oil and the 'capital-and-technology' oil. However, more serious competition, even for the largest state oil companies like PDVSA, Pemex and Pertamina is likely to result from governments' efforts to commercialize their operations and open the oil sector up to more private investment.[336] In Pertamina's case, plans to privatize it have been put before parliament; the fates of PDVSA and

[335] PIRINC, 'Market Factors Not Price Dumping -An Analysis of the 1998 Oil Price Collapse', *MEES*, vol. XLII, no. 32, 9 August 1999, p. 8. (http://www.mees.com/dotcom/current/42n32d01.htm)

[336] News release, Energy Intelligence Group, 10 June 1998, p. 1. (http://www.piwpubs.com/prpiwtop.htm)

Pemex are still debated[337]. The margin for inefficiencies is becoming very small indeed. The state oil companies came under severe pressure in 1998-1999, when oil prices neared the $10 a barrel threshold. Although, in March 1999 OPEC member states again agreed to reduce output and prices responded to this reduction in output by doubling from around $10 a barrel to around $21 a barrel in August 1999, the temptation to cheat is potentially still very large. The success of the March 1999 output restrictions can only be sustained if OPEC manages to adjust production levels to demand smoothly. Any over- or undershooting will create price volatility and the temptation to cheat. However, such fine-tuning may be made more difficult due to the lack of accurate information on production and stock levels of crude oil and oil products and a proper evaluation of demand. In addition, OPEC will need to consider how Iraq can be integrated into OPEC production policies when the UN regulates foreign direct investments into the Iraqi oil industry as a step towards rehabilitation.[338] Despite the resolve shown in March 1999, the underlying instability of OPEC is too great to function successfully.

8.7 Merger Mania

The 1998-1999 price collapse exposed not only the vulnerabilities of the 'higher-extraction-costs' state oil companies but also those of the international oil companies. The price collapse provoked an acceleration in industry and firm restructuring, which all parties tried to achieve both through increased cost-cutting and through mergers and takeovers. As long as the low extraction cost oil in the Middle East was not open to private international oil company investments, the international oil companies had to establish their competitive edge on the basis of organizational, financial, human capital and technological efficiencies. The inefficiency of state oil companies and the dependence of governments on oil revenues were always an incentive to maintain oil prices above a certain level. That is why Saudi Arabia, for instance, always maintained a target price of $14-18 a barrel. If prices fell below this level, OPEC countries were compelled to reduce output and support prices. However, this stop-go policy has now come under pressure because the market no longer clears at this price level,

[337] See Chapters 3 and 7.

[338] *Financial Times*, 16 August 1999.

and more oil than in the previous decade, is being produced by private international oil companies.

Price volatility as much as the price level has encouraged the international oil companies to sharpen their investment criteria and performance goals. At Shell, for instance, investments in upstream oil developments, must generate a proper return on investments at a price of $10 a barrel.[339] Such a policy on the part of the large international oil companies increases the pressure to restructure on the oil-producing countries whose extraction costs are in the $7-11 range. Their competitive edge will vanish unless they start making their organization more efficient. Given the state of the economy in these countries (Venezuela, Nigeria, Algeria, Indonesia and Iran for instance), the state oil company has little to expect from the government. The state oil companies will be unable to compete with the private international oil companies unless the oil and gas regime are changed. The yardstick for attracting capital into oil developments will be based on the investment criteria of the international oil companies. Non-compliance with these criteria will result either in no capital or in steep mark-ups on interest rates, which the oil-producing countries can ill afford.

Another complicating factor is that the rehabilitation of Iraq on the international oil market will open up this low-extraction cost oil producer for foreign direct investments. Such a development will also challenge the position of state oil companies in other low-extraction-cost countries. Despite the accomplishments of the March 1999 output reductions, the oil price collapse in 1998/99 has transformed the conditions of competition. No international oil company can afford to depend on OPEC production policies to sustain its business. They have therefore sharpened the yardstick and are seeking continuity in merging their interests with other oil companies.

The mega-mergers of Exxon-Mobil, BP Amoco and Arco, Total-Petrofina, and the bid of TotalFina for Elf Aquitaine of France have upset the organization of the international oil market. The massive move to on the part of the largest international oil companies to consolidate and reap synergy benefits must eventually result in large cost reductions for the new companies. The estimated cost reduction at Exxon/Mobil, for instance, is

[339] *Beleggers Belangen*, no. 7, 19 February 1999, p. 10.

$4 billion.[340] The optimism regarding potential synergy benefits in Exxon-Mobil's case was significant because the assets of the two companies overlap only slightly, and the earnings of Exxon would be enhanced, despite the premium of 15% paid over the share price of Mobil.[341]

The synergy gains in the BP Amoco merger were also apparent, unlike the new group's subsequent takeover of Arco, where the synergy gains were regarded as less obvious and where BP Amoco ended up paying 20% above the share-value of the company. BP Amoco and Arco hope to realize cost reductions similar in scale to those achieved at Exxon/Mobil. The BP Amoco merger alone resulted in the scrapping of nearly 8,000 jobs. The group further announced that it intends to sell $10 billion in assets (one third of the group's refining capacity, nearly $4 billion of its exploration and production assets, and $2.5 billion of its chemical operations), and invest $26 billion over the next three years to meet new financial targets, and to arrive at a finding, developing and producing cost base of $5-6 a barrel.[342] The total restructuring costs are estimated to be $2.5 billion over two years.[343] The merger strategy of BP and the subsequent restructuring plans of BP Amoco/Arco constituted an aggressive departure from its earlier strategy. The rate of return on BP Amoco's investments was much lower than that achieved by its new closest rivals Exxon and Shell. The combination of cost reductions, asset sales and new investments is intended to upgrade BP Amoco's portfolio and lead to a balance in international operations similar that to which Exxon and Shell already have. However, the integration of two companies into BP also carries a lot of risks. The scale of the operation is very large, and if the performance criteria are not met, for whatever reason, the benefits will be in jeopardy.

The danger of a period of extensive merger and takeover activity is that other companies may be compelled to join in (herding) in order to prevent other companies from acquiring an already identified target firm. They may also join in the merger and takeover activity to prevent an asset falling into the hands of a rival firm, a situation which would threaten their market position. As merger and takeover activity increases, the targets that best fit

[340] *Financial Times*, 28 November 1999.

[341] *Financial Times*, 28 November 1999.

[342] *Financial Times*, 19 July 1999.

[343] *Het FIA Beleggersweekblad*, 21 April 1999, no. 16, p. 6.

in with the activities of the acquiring firm have been snapped up, and second best options are pursued. This could entail greater restructuring costs and put into doubt the efficiency of the merger or takeover. In such cases, mergers and takeovers change from being value maximizing to non-value maximizing exercises. As Schenk notes:"Rather the existence of strategic interdependence under uncertainty, conditioned by the availability of funds, may compel managements to undertake mergers even if these will not increase economic performance. Inertia may prevail for long periods, but as soon as an initial, clearly observable move has been made by one of the major players, it is likely that other players will rapidly follow with similar moves. With multi-market oligopoly omnipresent, and given the increasing weight assigned to stock market performance appraisals, the ultimate result may be an economy-wide merger boom."[344]

Compared to the consolidation of Exxon and BP, that of Shell seems rather sedate. Rather than joining the merger mania, Shell's efforts are geared towards achieving internal cost reductions of $2.5 billion and restructuring the internal organization. In terms of market value, Exxon-Mobil is now the largest company with a market value of $270 billion, followed by BP Amoco-Arco at $200 billion.[345] Shell's market value is estimated to be at $190 billion. The resistance company's, thus far, to engaging in a substantive merger or takeover is based on its believe that it is still large enough to compete successfully and that internal growth will generate a superior return on assets.[346] In this opinion Shell is supported by certain literature on mergers and acquisitions, which concludes that not only did the acquirer's rate of return decline but also that there was a permanent decline in profitability.

Independent US oil producers are also being swept up in the merger activity. Many companies' high debt ratios and full-cost accounting will force them to re-evaluate their reserves at end-1998 commodity prices, and this could result in defaults. Other companies' share prices are so low that their oil reserves are valued at $1-$3 a barrel, which makes it cheaper to

[344] H. Schenk, *op. cit.*, Management report no. 21-1999, p. 11.

[345] *Financial Times*, 16 August 1999.

[346] *NRC Handelsblad*, 5 June 1999, Interview with M. Van den Bergh, President-Director of Koninklijke Olie.

buy than to find and develop oil reserves.[347] The general trend in this period of merger activity is to consolidate the assets in mature oil (or oil product) markets like Europe and the US, to increase international gas assets, and to invest in new oil exploration and production assets under stringent performance criteria that meet the state oil companies head on.

8.8 CONCLUSION

The international oil companies are now facing a completely different international oil market than in 1973, when their main reserve and production assets were nationalized. The break-up of the value chain is slowly being restored as more state oil companies are being privatized. The large international oil companies, engaged in a consolidation process, are able to compete as a result of investment strategies and the development of their financial, human capital and technological advantages. They were initially able to develop their present day competitive edge under the price umbrella of OPEC. After 1986, prices declined to a price band of around \$15-\$18, high enough for the oil companies to sustain their 'high-extraction-cost' oil. The availability of new exploration and production areas, particularly after the break-up of the Soviet Union, signalled a new round of competition, this time not only among the private oil companies but also among the private and state oil companies. The decrepit state of public finances in many oil-producing countries, including OPEC member states together with the competition for investment capital among oil-producing countries has resulted in a substantial change in the oil regimes in producing countries. A move towards liberalization and privatization has become the rule rather than the exception. The international oil companies are ready to reap the benefits of the oil-producing countries' experiment with state ownership. The re-integration of 'low-extraction-cost' oil into the investment market will modernize the sector further. However, a successful transformation can only take place within a proper regulation regime, in which the government can collect its taxes, and the environment is protected against degradation. It is the oil-producing countries' task to make sure that the reconstruction of the oil and gas regime is not carried out at the 'gun point' of further macro-economic instability. They rather need to preempt such a situation and act while there is still value stored in their state oil companies.

[347] *International Petroleum Finance*, January 1999, p. 1.

CHAPTER 9

STRATEGIC CONSOLIDATION

The days of frantic activity surrounding OPEC's Ministerial Meetings are over. Speculation on new oil-price levels have been replaced by speculation on who will break an output agreement first. Yet it is unfair and unwise to wallow in scepticism. OPEC has matured into an important platform for discussion among developing oil-producing countries with differing economic and political interests. The output restrictions of March 1999 not only came to the rescue of the member states themselves, but also gave other players in the industry a breathing space after months of very low oil prices. The member states, in particular Saudi Arabia, are again forfeiting market share for the stability of the international oil market. Moreover, they are doing that in the knowledge that, within a couple of months, other players in the market will have increased production and reap additional revenues. Despite commanding of three-quarters of the world's proven oil reserves, OPEC member states have to accept that they can only produce 42% of world production. The March 1999 agreement therefore provides nothing more than a temporary respite.

The dominant position of the oil-producing countries is being challenged in the international oil market by the large international oil companies. The large private international oil companies, already comparable in size to the largest of the NOCs, have managed to make significant advances in efficiency in the field of organization, finance, human capital and technology. Their ability to compete has improved dramatically since 1973. Moreover, the availability of new opportunities for exploration in the former Soviet Union has allowed the private international oil companies to expand their activities and lure capital away from the OPEC countries. As a result, the average cost of a barrel of oil produced by the private companies has decreased substantially since the 1970s. At the same time, costs for the NOCs have increased due to organizational inefficiencies. The difference between 'low-cost-extraction' oil and 'high-cost-extraction' oil, the critical element in the competitive advantage of the OPEC countries, has been significantly narrowed. Unless the oil-producing countries acknowledge the root of their problems, they will end up surrendering their remaining competitive advantage by clinging to an obsolete oil regime.

The underlying dynamic changes in the market are continuing and require policy responses from the member states that go much further than endlessly repeating inadequate (short-term) policies, in the hope that this time they will work. The agreements to restrict output are stop-gap policies, which fight the symptoms rather than the disease. The oil-producing countries have to rethink the role of the state in their economies, and the role they have assigned to their NOCs. Neither OPEC nor the NOCs have able to provide the oil-producing countries with a stable level of economic rents. Rather, the structure of the domestic oil sector and OPEC's policies have created a boom-bust cycle economy, in which the fluctuations in the economy are more severe than the oil cycle itself. Management of the economy becomes increasingly complicated because of the continuous over- and undershooting. Macro-economic instability becomes the rule rather than the exception. The upswings in the oil market are almost as bad as the downswings, because each time oil revenues recover, efforts to restructure the economy slack off.

My conclusion is that oil-producing countries should immediately suspend their stop-go policies and make a serious effort to break the downward spiral. The only way to implement a restructuring of the economy is to reorganize the oil sector. With its deadly myopic policies, the oil sector has taken the entire economy hostage.

The state should withdraw from the production of oil, privatize the NOCs and collect economic rents through taxation. Such a regime will prevent the misuse of the NOCs for purposes other than the efficient production of oil (and gas). Privatization and international and domestic competition will stimulate the efficiency of the NOCs. If they are unable to compete, they should be allowed to go bankrupt or other more efficient firms will take them over. In a liberalized oil and gas regime, the government should concentrate on monitoring the sector through a new regulatory regime. Government policy goals, such as sustaining the exploration and production of oil and gas, securing economic rents, protecting the environment, prudent depletion policies, de-monopolize crude oil reserves, developing competitive bidding schemes, and so on should be integrated into the regime. Moreover, general economic regulation and economic legislation (on bankruptcy, competition, banking and finance) should underpin the oil regime. The information problem should be addressed in strict disclosure rules to safeguard governments' share of economic rents and other policy interests.

Governments must learn to make their economic policies less dependent on oil revenues. Only a sustainable level of oil revenues should be included in their fiscal policies. The oil-producing countries' first priority should be the stabilization of the domestic oil sector and the overall economy. The liberalization of the economy and privatization of the oil sector should be important elements in such a strategy.

Governments must provide regulatory transparency and set-up proper monitoring institutions, such as a competition board, independent of the government. In order to break with the extensive culture of corruption in many oil-producing countries, the competition board should be given the power to investigate companies at any time it likes, and bring in accountancy firms from abroad if necessary to carry out additional research. For reasons of transparency, the board should publish its findings.

The drastic changes that need to be made in the oil-producing countries' regulatory regimes will not happen overnight. It is more likely that the majority of the countries will continue to muddle on, and hope for better times. The international oil companies will continue to advance, and most probably will proceed to expand from country to country as each opens up. If that happens, the first-mover countries will be able to benefit at the expense of the others. This is particularly important for 'higher-extraction-cost' countries, whose competitive advantage is already rather slim. These countries can no longer afford to wait. Saudi Arabia and the other countries in the Gulf will play a key part in the outcome of this process of change. If they elect to be among the first-mover states, the other countries will be under huge pressure to follow, and consequently under less fortunate circumstances.

Saudi Arabia and Kuwait in particular, or more precisely Saudi Aramco and KPC, could join the consolidation game of the international oil companies and make the complete or partial privatization of their NOCs an integral part of the takeover of a large international oil company. Such an aggressive method of privatization would enhance their internationalization strategy and challenge their main competitors. Governments can force efficiency on their state oil companies (whether partially or wholly state-owned) if they privatize them with smaller exploration acreage rights than they have at present, and organize competitive bidding for these exploration areas. The introduction of competitive bidding for exploration and production rights will create the political and economic climate necessary to convince other countries' 'Monopolies and Merger Commissions' or competition boards, particularly in the acquired firm's home country, to accept such a deal. As long as the state remains the prime shareholder, the potential for international expansion will be limited. Introducing a new business climate in the international oil industry will not change the strategic importance of oil. If producing governments change their paths to liberalization and privatization, they must implement these changes in ways that will not encourage political intervention on the part of other states.

References

Adelman, M., Crude Oil Production Costs (1966), in: *The Economics of Petroleum Supply*, MIT Press, Cambridge Mass., 1993.

Adelman, M., Efficiency Of Resource Use in Crude Petroleum (1964), in: *The Economics of Petroleum Supply*, MIT Press, Cambridge, Mass., 1993.

Adelman, M., Crude Oil Production Costs (1966), in: *The Economics of Petroleum Supply*, MIT Press, Cambridge Mass., 1993.

Al-Yousuf, A., *Kuwait and Saudi Arabia: From Prosperity to Retrenchment*, O.I.E.S. papers on Oil and Finance 7, Oxford Institute for Energy Studies, Oxford, 1990.

Anastassopoulos, J-P., Blanc, G., and Dussauge, P., *State-owned Multinationals*, Wiley/IRM series on Multinationals, Wiley & Sons, Chichester, 1987.

Armstrong, M., Cowan, S., and Vickers, J., *Regulatory Reform, Economic Analysis and British Experience*, The MIT Press, Cambridge Mass., 1994.

Atkinson, A.B., and Stiglitz, J.E., *Lectures on Public Economics*, McGraw-Hill International Editions, Economics Series, McGraw-Hill Book Company Ltd, London, 1980.

Auty, R.M., *Sustaining Development in Mineral Economies, The Resource Curse Thesis*, Routhledge, London/New York, 1993.

Auty, R.M., *Resource-Based Industrialization, Sowing the Oil in Eight Developing Countries*, Clarendon Press, Oxford, 1990.

Baldwin, R.E., 'Adapting the GATT to a more regionalised world: a political economy perspective'. In *Regional Integartion and the Global Trading System*, K. Anderson and R. Blackhurst (eds.), Harvester/Wheatsheaf, Hemel Hampstead, 1993.

Bardhan, P., Corruption and Development: A Review of Issues, *Journal of Economic Literature, vol. XXXV* (September 1997).

Barnes, P., *Indonesia, The Political Economy of Energy*, Oxford University Press, Oxford, 1995.

Barrows, G., A Survey of Incentives in Recent Petroleum Contracts, in: N. Beredjick and T. Wälde (eds.), *Petroleum Investment Policies in Developing Countries*, Graham & Trotman, London, 1988.

Baumol, W.J. and Blinder, A.S., *Economics, Principles and Policy*, Fifth edition, Harcourt Brace Jovanovich, Publishers, San Diego/New York, 1991.

Bentham, R., Legal Status of State Petroleum Companies, in: N. Beredjick and T. Wälde (eds.), *Petroleum Investment Policies in Developing Countries*, Graham & Trotman, London, 1988.

Bentham R.W., and Smith, W.G.R., *State Petroleum Corporations: Corporate Forms, Powers and Control*, Centre for Petroleum and Mineral Law Studies, University of Dundee, Scotland, 1986.

Bird, G. , *World Finance and Adjustment, An Agenda for Rreform*, Macmillan, Basingstoke, 1985.

Boers, J., Eisenloeffel, G., van der Linde, C., Olieraffinage en petrochemie in de Golfstaten en EG, in: *ESB*, 22 February 1984.

Bourgeois, B., and Rodriguez-Padilla, V., Oil Exploration and Production Between the 1980s and 1990s: Geological Advantage and International Petroleum Orders, *Energy Studies Review*, Vol. 3, no.3, 1991.

Bourrelier, P-H., Boy de la Tour, X., and Lacour, J-J., Energy in the Long term, Mobilization or laissez-faire?, *Energy Policy*, March 1992.

BP Amoco *Statistical Review of World Energy* 1999.

BP Statistical Review of World Energy, June 1990.

BP *Statistical Review of World Energy*, 1998.

Brennan M., and Schwarz, E., A New Approach to Evaluating Natural Resource Investments, in: D.H. Chew jr., *The New Corporate Finance, Where Theory Meets Practice*, McGraw-Hill, New York, 1993.

Camdessus, M., *Capital Account Liberalization and the Role of the Fund*, IMF, Washington D.C., March 9, 1998.

Casas Gonzalez, A., *How National Oil Companies are Meeting the New Conditions*, Montreux Energy Roundtable 1991, New Management and Investment Challenges 17-19 April 1991.

Cave, M., *Recent Developments in the Regulation of Former Nationalised Industries*, Government Economic Service Working Paper, No. 114/ Treasury Working paper No. 59, London, 1991.

Caves, R.E., Frankel, J.F., Jones, R.W., *World Trade and Payments, An Introduction*, HarperCollins College Publishers, New York, 1993.

Crystal, J., *Oil and Politics in the Gulf: Rulers and Merchants in Kuwait and Qatar*, Cambridge University Press, Cambridge, 1990.

Dam, K.W., *The Rules of the Game, Reform and Evolution in the International Monetary System*, The University of Chicago Press, Chicago/ London, 1982.

De Haan, W., Van der Molen, F., and Van der Linde, C., Een Zwaargewicht slankt af: Unilever 1965-1980, in: F. Crone and H. Overbeek, *Nederlands Kapitaal over de grenzen*, SUA, Amsterdam, 1981.

De Jong, H.W., Concurrentie en Concentratie in de Markteconomie, in: J.G.C.M. Bune, J.G. Groeneveld et.al.(eds.), *Actualiteiten en Accountancy '91*, Nivra, Samsom Bedrijfsinformatie, Alphen aan den Rijn, 1991.

De Jong, H.W., *Dynamische Markttheorie*, Bedrijfseconomische Monografieën, Stenfert Kroese, Leiden/Antwerpen, 1985.

De Jong, H.W., The State as Entrepreneur, The rise and fall of an idea in the economic experiences of European countries, in: G. Aschinger et. al., *Deregulierung- eine Herausforderung an die Wirtschafts- und Sozialpolitik in der Marktwirtschaft*, Duncker & Humblot, Berlin, 1989.

Dornbusch, R., *Policymaking in the Open Economy, Concepts and Case Studies in Economic Performance*, Oxford University Press, Oxford, 1993.

Dornbusch, R., 'Debt Problems and the World Macroeconomy'. In *Developing Country and the World Economy*, J.D. Sachs, NBER, The University of Chicago Press, Chicago, 1989.

Edwards, C., *The Fragmented World, Competing Perspectives on Trade, Money and Crisis*, Methuen, London, 1985.

Eichengreen , B., and Kenen, P.B., Managing the World Economy under the Bretton Woods System: An Overview, in: P.B. Kenen (ed.), *Managing the World Economy, Fifty Years after Bretton Woods*, Institute for International Economics, 1994.

Eichengreen, B., *Globalizing Capital, A history of the International Monetary System*, Leuven University Press, Leuven, 1996.

Eichengreen, B., *International Monetary Arrangements for the 21st Century*, The Brookings Institution, Washington D.C., 1995.

Elster, J., *The Cement of Society. A Study of Social Order*, Cambridge University Press, Cambridge, 1989.

Elting Treat, J., *National Oil Companies: Challenges for the Future*, Booz, Allen & Hamilton Inc., Middle East Petroleum and Gas Conference, Dubai, January 13, 1993.

Energy Economist, December 1991.

Energy Intelligence Group, Privatization Changing the Face of the World Oil Industry According to Ranking the World's Top Oil Companies, June 10, 1998 and December 1998.

Energy Intelligence Group, *Company News*, 15 December 1998.

Energy Intelligence Group, News release, 10 June 1998.

Energy Intelligence Group, What BP Amoco sees in Arco, Industry News,, March 29, 1999.

Evans, J., *OPEC, Its Member States and the World Energy Market*, A Keesing Reference publication, Longman, Harlow, 1986.

Fee, D., *Petroleum Exploitation Strategy*, Belhaven Press, London/New York, 1988.

Financial Times, 16 August 1999.

Financial Times, 30 March 1999.

Financial Times, 19 July 1999.

Financial Times, 10 March 1999.

Financial Times, 28 November 1998.

Financial Times, 13 July 1999.

Financial Times, 17 August 1999.

Financial Times, 16 August 1999.

Financial Times, 26 February 1999.

Financial Times, 12 May 1999.

Finon, D., 'The Prospects for a new International Petroleum Order', *Energy Studies Review*, vol.3, no.3, 1991.

Fishlow, A., Inequality between Nations, in: *Economic Development*, J. Eatwell, M.

Friedman, I.S., The International Monetary Fund: A Founder's Evaluation. In Myers, R.J., *The Political Morality of The International Monetary Fund*, Ethics and Foreign Policy, volume 3, Transaction Books, New Brunswick, N.J., 1987.

Gavin, M., 'Adjusting to a Terms of Trade Shock: Nigeria, 1972-88'. In Dornbusch, R., *Policymaking in the Open Economy, Concepts and Case Studies in Economic Performance*, Oxford University Press, Oxford, 1993.

Higgott, R., *Review of Globalisation*, paper prepared for The Economic and Social research Council, 5 June 1998.

Indonesia, *Draft Law of the Republic of Indonesia concerning Natural Oil and Gas*, second draft 10 February 1999.

Milgate, P. Newman (eds.), The New Palgrave, The Macmillan Press Ltd., London, 1987.

Myers, R.J., *The Political Morality of The International Monetary Fund*, Ethics and Foreign Policy, volume 3, Transaction Books, New Brunswick, N.J., 1987.

Frankel, P., Essentials of Petroleum: A Key to Oil Economics, in: I. Skeet (ed.), *Paul Frankel Common Carrier of Common Sense, a selection of his writings 1946-1988*, Oxford: OIES, 1989.

Garretson de Vries, M., *Balance of Payments Adjustment, 1945 to 1986, The IMF Experience*, International Monetary Fund, Washington D.C., 1987.

Gelb, A. and Associates, *Oil Windfalls, Blessing or Curse*, A *World Bank* Research Publication, Oxford University Press, Oxford, 1988..

Gelb, A., The Oil Syndrome: Adjustment to Oil Windfall Gains in Oil-Exporting countries, in: D. Lal and M. Wolf (eds.), *Stagflation, Savings and the State*, Oxford University Press, Oxford, 1986.

Grayson, L.E., *National Oil Companies*, John Wiley and Sons, Chichester, 1981.

Griffin J.M., and Steele, H.B., *Energy Economics and Policy*, Academic Press, New York.

Gustafson, T., *Crisis amid Plenty*, Princeton University Press, 1989.

Hartshorn, J.E., *Oil Trade: Politics and Prospects*, Cambridge University Press, Cambridge, 1993.

Helm, D., The Economic Border of the State, in: *The Economic Border of the State*, D. Helm (ed.), Clarendon Paperbacks, Oxford University Press, 1990.

Helm, D., and Jenkinson, T., Introducing Competition into Regulated Industries, in: D.Helm and T. Jenkinson (eds.), *Competition in Regulated Industries*, Oxford University Press, Oxford, 1998.

Het FIA Beleggersweekblad, 21 April 1999, no. 16.

Hettne, B., *Development Theory and the Three Worlds*, Longman Development Studies, Longman Scientific & Technical, New York, 1990.

Holmstrom, B., The Firm as a Subeconomy, *Journal of Law, Economics & Organization*, vol. 15, no. 1, April 1999.

Horn, M.J., *The Political Economy of Public Administration*, Cambridge University Press, 1995.

Horsnell P., and Mabro, R., *Oil Markets and Prices, The Brent Market and the formation of world oil Prices*, Oxford University Press, Oxford, 1993.

Institute of International Finance, *Annual Report 1997*.

International Petroleum Finance, January 1999.

Jacquemin, A., *The New Industrial Organization, Market Forces and Strategic Behaviour*, Clarendon Press, Oxford, 1987.

Jensen, M.C., The Modern Industrial revolution, Exit, and the Failure of Internal Control Systems, *The Journal of Finance*, vol.XLVIII, no. 3, July 1993.

Karl, T.J., *The Paradox of Plenty, Oil Booms and Petro-States*, University of California Press, Berkeley, 1997.

Karshenas, *M., Oil, State and Industrialization in Iran*, Cambridge University Press, Cambridge, 1990.

Kemp, A.G., Petroleum Policy Issues in Developing Countries, *Energy Policy*, February 1992.

Kensinger, J.W., and Martin, J.D., Project Finance: Raising Money the Old-Fashioned way, in: D.H. Chew jr., *The New Corporate Finance, Where Theory Meets Practice*, McGraw-Hill, New York, 1993.

Kindleberger, C. P., *A Financial History of Europe*, Oxford University Press, Oxford, 1993.

Krugman, P. R., *Currencies and Crises*, The MIT Press, Cambridge, Mass., 1993.

Krugman, P. R., *Peddling Prosperity, Economic Sense and Nonsense in the Age of Diminished Expectations*, W.W. Norton & Company, New York, 1994.

Krugman, P. R., and Obstfeld, M., *International Economics, Theory and Policy*, Addison-Wesley, Reading, Mass., fourth edition, 1997.

Kumar Bagchi, A., Industrialization, in: *Economic Development*, J. Eatwell, M. Milgate, P. Newman (eds.), The New Palgrave, The Macmillan Press Ltd., London, 1987.

Laffont J-J. and Tirole, J., *A Theory of Incentives in Procurement and Regulation*, The MIT Press, Cambridge Mass., 1993.

Laffont, J-J., and Tirole, J., *Creating Competition through Interconnection: Theory and Practice*, mimeo, 1994.

Leite C., and Weidmann, J., *Does Mother Nature Corrupt? Natural Resources, Corruption, and Economic Growth*, IMF Working Paper, WP/99/85.

Mabro, R., et al, *The Economics of North Sea Oil*, Oxford, OIES/OUP, 1986.

Mabro, R., The Changing Nature of the Oil Market and OPEC Policies, *OPEC Review*, vol. VI, no. 4, Winter 1982.

Macey, J., and Miller,G., Corporate Governance and Commercial banking: A Comparative Examination of Germany, Japan, and the United states, *Stanford Law Review* 48, (1995).

Machmud, T.N., *The Production Sharing Contract in the Indonesian Petroleum Industry: at a crossroads after 30 years of history...... status quo or reform?*, doctoral thesis, Fall 1999, Leiden University.

McGowan,F., The Development of Orimulsion and Venezuelan Oil Strategy, *Energy Policy*, March 1992.

Messer, M., *The Reorganization of the Russian Oil Industry*, Draft Doctoral Thesis, Leiden, to be completed 2000.

Miami Herald, 6 August 1999.

Milgrom P., and Roberts, J., *Economics, Organization and Management*, Prentice Hall International, Edgewood Cliffs, 1992.

Mnookin, R.H., and Wilson, R.B., Rational Bargaining and Market Efficiency: Understanding Pennzoil v. Texaco, *Virginia Law Review*, vol. 75, 1989.

Moran, T.H., *Oil Prices and the Future of OPEC. The political economy of tension and stability in the Organization of Petroleum Exporting Countries*, Research Paper R-8, resources for the Future, Washington D.C., 1978.

Murphy, K.M., Shleifer A. and. Vishny, R.W, Why is rent Seeking so Costly to Growth?, Chapter 4, in: A. Shleifer and R.W. Vishny, *The Grabbing Hand, government pathologies and their cures*, Harvard University Press, Cambridge Mass. 1998.

Newbery, D., Taxation and Development, in: *The Theory of Taxation in Developing Countries*, D. Newbery and N. Stern (eds.), World Bank/Oxford University Press, Oxford, 1987.

NRC Handelsblad, 5 June 1999, Interview with M. Van den Bergh, President-Director of Koninklijke Olie.

Obstfeld, M., and Rogoff, K., *Foundations of International Macroeconomics*, MIT Press, Cambridge, Mass., 1997.

Olorunfemi, M.A., The Dynamics of national oil companies, *OPEC Review*, Winter 1991, volume xv, no.4.

OPEC, *Facts & Figures, a graphical analysis of world energy up to 1985*, 1986.

OPEC, *Official Resolutions and Press Releases*, 1960-1990, The Secretariat OPEC, Vienna.

OPEC *Annual Statistical Bulletin* 1991.

Ordeshook, P.C., The Emerging discipline of political economy, in: J.E. Alt and K.A. Shepsle (eds.), *Perspectives in Positive Political Economy*, Cambridge University Press, Cambridge, 1990.

Petroleum Economist, December 1991.

Petroleum Economist, January 1992.

Petroleum Economist, Nigeria: a special Report, March 1992.

Petroleum Economist, April 1992.

Petroleum Economist, April 1975.

Petzinger, T., jr., *Oil & Honor, The Texaco-Pennzoil Wars, inside the $11 billion Battle for Getty Oil*, G.P. Putnam's Sons, New York, 1987.

PIRINC, Market Factors Not Price Dumping- An Analysis of The 1998 Oil Price Collapse, *MEES*, vol. XLII, no. 32, 9 August 1999.

Pollio, *G., Financing Energy Development In An Uncertain World, The Role Of Project Finance*, Presentation, Faculty of Law, Leiden University, Spring 1993.

Qian, Y and Xu, C., *Innovation and Bureaucracy Under Hard and Soft Budget Constraints*, Review of Economic Studies, 1998.

Randall, L., *The Political Economy of Venezuelan Oil*, Praeger, New York, 1987.

Rasmusen, E., *Games and Information, An Introduction to Game Theory*, Basil Blackwell, Oxford, 1989.

Reve, T., The Firm as a Nexus of Internal and External Contract, in: *The Firm as a Nexus of Treaties*, Masahiko Aoki, Bo Gustafsson and Oliver E. Williamson (eds.),Sage Publications, London, 1990.

Richard, J., Can OPEC Make it Stick?, *Barron's*, May 17, 1999.

Roeber, J., Oil Industry Structure and Evolving Markets, *The Energy Journal*, Special Issue, 1994.

Roe Goddard, C., Passe-Smith, J.T. , Conlin, J.G., *International Political Economy, State-Market Relations in the Changing Global Order*, Lynne Rienner Publishers, Boulder, 1996.

Romano, R., A Guide to takeovers: Theory, Evidence, and Regulation, *The Yale Journal on Regulation*, vol. 9, 1992.

Sachs, J.D., Conditionality, 'Debt relief, and the Developing Country Debt Crisis'. In *Developing Country and the World Economy*, J.D. Sachs, NBER, The University of Chicago Press, Chicago, 1989.

Sampson, A., *The Seven Sisters*, Coronet Books, Sevenoaks, 1975.

Santos, A., Privatization and State Intervention- (An Economic Approach), in: *Deregulation or Re-regulation?, Regulatory reform in Europe and the United States*, Giandomenico Majone (ed.), Pinter Publishers, London, 1990.

Saudi Arabian Monetary Agency, *Thirty-Fourth Annual Report*, 1419H (1998G).

Schenk, H., *Policy Implications of Purely Strategic Mergers*, Erasmus University Management Report no. 21-1999.

Scott, H.S., Wellons, P.A., *International Finance, Transactions, Policy, and Regulation*, University Casebook Series, Foundation Press, New York, 1998.

Seymour, A., *Refining and reformulation: the challenge of green motor fuels*, Oxford Institute for Energy Studies, Oxford, 1992.

Shleifer, A and Vishny, R.W., *Politicians and Firms*, Discussion Paper Number 1686, Harvard Institute of Economic Research, June 1994.

Shleifer, A., State versus Private Ownership, *Journal of Economic Perspectives*, Fall 1998, vol. 12, no.4.

Shleifer A., and Vishny, R.W., *The Grabbing Hand, Government Pathologies and their cures*, Harvard University Press, Cambridge, Mass., 1998.

Shwadran, B., *The Middle East, Oil and the Great Powers*, Atlantic Press, 1955.

Siegel, D., Smith, J., and Paddock, J., Valuing Offshore Oil properties with option Pricing Models, in: D.H. Chew jr., *The New Corporate Finance, Where Theory Meets Practice*, McGraw-Hill, New York, 1993.

Smith, C.W., Smithson, C.W., Wilford, D.S., Managing Financial Risk, in: D.H. Chew jr., *The New Corporate Finance, Where Theory Meets Practice*, McGraw-Hill, New York, 1993.

Stiglitz, J.E., *Whither Socialism*, MIT Press, Cambridge, Mass, 1995.

Strange, S., *States and Markets, An Introduction to International Political Economy*, Pinter Publishers, London, 1988.

Subruto, Dr., How OPEC perceives the energy outlook for this decade, *OPEC Bulletin*, April 1992.

The Economist, Cheap Oil, The next shock?, March 6th 1999.

The Economist Intelligence Unit, Country Profile 1989-90 Nigeria.

The Economist Intelligence Unit, Country profile Nigeria 1993/94.

Tirole, J., A Theory of Collective Reputations (with applications to the persistence of corruption), *Journal of Economic Studies*, vol. 63 (1996).

Tussie, D., *The Less Developed Countries and the World Trading System, A Challenge to the Gatt*, Frances Pinter, London, 1987.

United Nations, United Nations Centre on Transnational Corporations, *Financial and Fiscal Aspects of Petroleum Exploitation*, UNCTC Advisory Studies, Series B, no. 3, ST/CTC/SER.B/3, New York, 1987.

Van der Linde, C., OPEC Adrift, Resources Policy 1994 20 (2).

Van der Linde, C., De olie-achtergrond van de Golfoorlog, *Armex Defensie Magazine*, January 1992.

Van der Linde, *C., Dynamic International Oil Markets, Oil Market Developments and Structure 1860-1990*, Kluwer Academic Publishers, Dordrecht, 1991.

Van der Linde, C. , Hervorming van het IMF, *Internationale Spectator*, jaargang 52, no. 11, november 1998.

Van der Linde, C., Empty Barrels Make the Most Sound, *World Competition*, Vol. 15, no. 2, December 1991.

Van Wijnbergen, S., 'The Dutch Disease': A Disease afterall?, *Economic Journal 94*, March, 1984; A. Gelb et. al., *op. cit*., 1988.

Vickers, J. and Yarrow, G., *Privatization: An Economic Analysis*, MIT Press, Cambridge, Mass., 1988.

Wade, R., *Governing the Market, Economic Theory and the Role of Government in East Asian Industrialization*, Princeton University Press, Princeton, 1990.

Wälde, T., Investment Policies in the International Petroleum Industry; responses to the current crisis, in: N. Beredjick and T. Wälde (eds.), *Petroleum Investment Policies in Developing Countries*, Graham & Trotman, London/ Dordrecht/Boston, 1988.

Wallerstein, I., Periphery, in: *Economic Development*, J. Eatwell, M. Milgate, P. Newman (eds.), The New Palgrave, The Macmillan Press Ltd., London, 1987.

Williams, M., End of An Era, *Petroleum Economist*, September 1991.

World Bank, The State in a Changing World, World Development Report 1997, *World Bank*/OUP, 1997.

Williamson, J., and Milner, C., *The World Economy, A Textbook in International Economics*, Harvester/Wheatsheaf, New York, 1991.

Yergin, D., *The Prize*, Simon & Schuster, New York, 1991.

Zonis, M., The New "Oil Shock" Political Instability is Next, *PIW*, February 1, 1999.

INDEX

Studies in Industrial Organization

1. H.W. de Jong (ed.): *The Structure of European Industry.* 1981
   2nd Revised edition, 1988: see below under Volume 8
2. M. Fennema: *International Networks of Banks and Industry* (1970–1976). 1982
   ISBN 90-247-2620-4
3. P. Bianchi: *Public and Private Control in Mass Product Industry.* The Cement Industry Cases. 1982 ISBN 90-247-2603-4
4. W. Kingston: *The Political Economy of Innovation.* (1984) 1989 2nd printing
   ISBN 90-247-2621-2
5. J. Pelkmans: *Market Integration in the European Community.* 1984
   ISBN Hb: 90-247-2978-5; Pb: 90-247-2988-2
6. H.W. de Jong and W.G. Shepherd (eds.): *Mainstreams in Industrial Organization.* 1986
   Book I: *Theory and International Aspects.* ISBN 90-247-3461-4
   Book II: *Policies: Antitrust, Deregulation and Industrial.* ISBN 90-247-3462-2
   Set ISBN Book I + II: 90-247-3363-4
7. S. Faltas: *Arms Markets and Armament Policy.* The Changing Structure of Naval Industries in Western Europe. 1986 ISBN 90-247-3406-1
8. H.W. de Jong (ed.): *The Structure of European Industry.* 2nd revised ed. (of Volume 1). 1988 Third Revised Edition, 1993: See below under Volume 18
9. I.L.O. Schmidt and J.B. Rittaler: *A Critical Evaluation of the Chicago School of Antitrust Analysis.* 1989 ISBN 90-247-3792-3
10. B. Carlsson (ed.): *Industrial Dynamics.* Technological, Organizational, and Structural Changes in Industries and Firms. 1989 ISBN 0-7923-9044-X
11. Z.J. Acs and D.B. Audretsch (eds.): *The Economics of Small Firms.* A European Challenge. 1990 ISBN 0-7923-0484-5
12. W. Kingston: *Innovation, Creativity and Law.* 1990 ISBN 0-7923-0348-2
13. B. Dankbaar, J. Groenewegen and H. Schenk (eds.): *Perspectives in Industrial Organization.* 1990 ISBN 0-7923-0814-X
14. P. de Wolf (ed.): *Competition in Europe.* Essays in Honour of Henk W. de Jong. 1991 ISBN 0-7923-1050-0
15. C. van der Linde: *Dynamic International Oil Markets.* Oil Market Developments and Structure (1860–1990). 1991 ISBN 0-7923-1478-6
16. D.B. Audretsch and J.J. Siegfried (eds.): *Empirical Studies in Industrial Organization.* Essays in Honor of Leonard W. Weiss. 1992 ISBN 0-7923-1806-4
17. R.J. Gilbert (ed.): *The Environment of Oil.* 1992 ISBN 0-7923-9287-6
18. H.W. de Jong (ed.): *The Structure of European Industry.* Third Revised Edition. 1993 ISBN Hb: 0-7923-2160-X; Pb: 0-7923-2253-3
19. D.K. Round (ed.): *The Australian Trade Practices Act 1974.* Proscriptions and Prescriptions for a More Competitive Economy. 1994 ISBN 0-7923-3228-8

Studies in Industrial Organization

20. A. van Witteloostuijn (ed.): *Market Evolution*. Competition and Cooperation. 1995 ISBN 0-7923-3350-0
21. G. Mussati: *Mergers, Markets and Public Policy*. 1995 ISBN 0-7923-3643-7
22. M.A. Jamison: *Industry Structure and Pricing*. The New Rivalry in Infrastructure. 1999 ISBN 0-7923-7708-7
23. C. van der Linde: *The State and the International Oil Market*. 1999 ISBN 0-7923-7709-5

Kluwer Academic Publishers – Dordrecht / Boston / London

MIX
Papier aus verantwortungsvollen Quellen
Paper from responsible sources
FSC® C105338

If you have any concerns about our products,
you can contact us on
ProductSafety@springernature.com

In case Publisher is established outside the EU,
the EU authorized representative is:
Springer Nature Customer Service Center GmbH
Europaplatz 3, 69115 Heidelberg, Germany

Printed by Libri Plureos GmbH
in Hamburg, Germany